# 3D Printing of Metals

# 3D Printing of Metals

Special Issue Editor

**Manoj Gupta**

MDPI • Basel • Beijing • Wuhan • Barcelona • Belgrade

**MDPI**

*Special Issue Editor*
Manoj Gupta
National University of Singapore
Singapore

*Editorial Office*
MDPI
St. Alban-Anlage 66
4052 Basel, Switzerland

This is a reprint of articles from the Special Issue published online in the open access journal *Applied Sciences* (ISSN 2076-3417) from 2017 to 2018 (available at: https://www.mdpi.com/journal/applsci/special_issues/3D_Metal_Printing).

For citation purposes, cite each article independently as indicated on the article page online and as indicated below:

LastName, A.A.; LastName, B.B.; LastName, C.C. Article Title. *Journal Name* **Year**, *Article Number*, Page Range.

**ISBN 978-3-03921-341-2 (Pbk)**
**ISBN 978-3-03921-342-9 (PDF)**

# Contents

# About the Special Issue Editor

**Manoj Gupta**, Dr., was a former head of the Materials Division of the Mechanical Engineering Department and director designate of the Materials Science and Engineering Initiative at NUS, Singapore. He received his Ph.D. from the University of California, Irvine, USA (1992), and was a postdoctoral researcher at the University of Alberta, Canada (1992). In August 2017, he was highlighted among the top 1% of scientists in the world by The Universal Scientific Education and Research Network and among the top 2.5% of scientists as per ResearchGate. He is credited with (i) the disintegrated melt deposition technique and (ii) the hybrid microwave sintering technique, an energy-efficient, solid-state processing method to synthesize alloys/micro/nanocomposites. He has published over 525 peer-reviewed journal papers and owns two US patents and one trade secret. His current h-index is 61, RG index is >47, and citations are greater than 14,000. He has also co-authored six books that have been published by John Wiley, Springer, and MRF, USA. He is Editor-in-Chief/Editor of twelve international peer-reviewed journals. In 2018 he was announced World Academy Championship winner in the area of Biomedical Sciences by the International Agency for Standards and Ratings. A multiple award winner, he actively collaborates with and visits Japan, France, Saudi Arabia, Qatar, China, USA, and India as a visiting researcher, professor, and chair professor.

*applied sciences*

**MDPI**

*Editorial*

# Special Issue: 3D Printing of Metals

**Manoj Gupta** ©

Department of Mechanical Engineering, National University of Singapore, 9 Engineering Drive 1,
Singapore 117576, Singapore; mpegm@nus.edu.sg

Received: 6 June 2019; Accepted: 11 June 2019; Published: 24 June 2019

Additive manufacturing (AM) has emerged as one of the most enabling new manufacturing technique; the topic has been extensively researched worldwide for almost two decades. The unique capabilities and potential of various AM techniques have led to almost homogeneous worldwide research efforts irrespective of international boundaries; such efforts have aimed at developing a thorough and critical understanding to harness the capabilities of AM that may translate to industrial practice. The motive behind these extensive research activities was to:

a. Optimize the use of materials to reduce wastage [1].
b. Optimize the use of manpower to enhance efficiency [1].
c. Optimize the use of resources to limit production time [2].

Both global governments and the private sector have invested billions of dollars to develop AM techniques to realize the goal of enabling sustainability as well as a profitable manufacturing route. All three categories of materials (metals/alloys, polymers, and ceramics) have been researched and practically all applications, whether in engineering or related to biomedical fields, have been equally targeted.

To further this cause, a Special Issue was launched in the MDPI journal 'Applied Sciences', which sought original contributions to develop further understanding of this fascinating area of manufacturing. A total of nine articles were accepted after a rigorous peer review process and subsequently published. Overall, the papers addressed:

a. AM process control/optimization including aspects of online monitoring [1,3,4]
b. Comparison studies with existing manufacturing methods to validate the acceptability of AM [5]
c. Product design and development [6]
d. Properties improvement using AM techniques [7]

Many AM techniques have been developed over last two decades; the work done thus far has enabled current researchers to understand both the scientific and technical capabilities and the limitations of these techniques. Accordingly, researchers have been clear in their selection of AM techniques, choices which have been primarily governed by material and end applications.

The articles collected in the present Special Issue indicate an emphasis on:

a. Metal-based materials including stainless steels, magnesium alloys, and nickel-based alloys [4,5,7].
b. Polymer-based materials [8,9].

The industrial sectors which are likely to benefit from the studies presented in this Special Issue include but not limited to the following sectors:

a. Construction
b. Transportation, including automobile and aerospace sectors
c. Nuclear
d. Biomedical

e.    Manufacturing

Articles presented in this issue are expected to be of considerable interest to students and researchers working in a wide spectrum of engineering and biomedical fields as well as for a number of existing and new applications.

Finally, I would like to thank all the authors for their excellent contributions to this Issue, to the reviewers for making useful comments to improve the quality of each article, and to the Applied Sciences editorial staff for processing and publishing these articles at their earliest convenience.

**References**

1.    Li, F.; Chen, S.; Shi, J.; Tian, H.; Zhao, Y. Evaluation and optimization of a hybrid manufacturing process combining wire arc additive manufacturing with milling for the fabrication of stiffened panels. *Appl. Sci.* **2017**, *7*, 1233. [CrossRef]
2.    Li, F.; Chen, S.; Shi, J.; Zhao, Y.; Tian, H. Thermoelectric cooling-aided bead geometry regulation in wire and arc–based additive manufacturing of thin–walled structures. *Appl. Sci.* **2018**, *8*, 207. [CrossRef]
3.    Chen, Z.; Zong, X.; Shi, J.; Zhang, X. Online monitoring based on temperature field features and prediction model for selective laser sintering process. *Appl. Sci.* **2018**, *8*, 2383. [CrossRef]
4.    Han, S.; Zielewski, M.; Martinez Holguin, D.; Michel Parra, M.; Kim, N. Optimization of AZ91D process and corrosion resistance using wire arc additive manufacturing. *Appl. Sci.* **2018**, *8*, 1306. [CrossRef]
5.    Manninen, M.; Hirvimäki, M.; Matilainen, V.; Salminen, A. Comparison of laser-engraved hole properties between cold-rolled and laser additive manufactured stainless steel sheets. *Appl. Sci.* **2017**, *7*, 913. [CrossRef]
6.    McEwen, I.; Cooper, D.; Warnett, J.; Kourra, N.; Williams, M.; Gibbons, G. Design & manufacture of a high-performance bicycle crank by additive manufacturing. *Appl. Sci.* **2018**, *8*, 1360. [CrossRef]
7.    Gao, Y.; Zhou, M. Superior mechanical behavior and fretting wear resistance of 3D-printed inconel 625 superalloy. *Appl. Sci.* **2018**, *8*, 2439. [CrossRef]
8.    Zhang, S. Degradation classification of 3D printing thermoplastics using fourier transform infrared spectroscopy and artificial neural networks. *Appl. Sci.* **2018**, *8*, 1224. [CrossRef]
9.    Jahan, S.; El-Mounayri, H. A thermomechanical analysis of conformal cooling channels in 3D printed plastic injection molds. *Appl. Sci.* **2018**, *8*, 2567. [CrossRef]

*applied*
*sciences*

MDPI

*Article*

# Evaluation and Optimization of a Hybrid Manufacturing Process Combining Wire Arc Additive Manufacturing with Milling for the Fabrication of Stiffened Panels

Fang Li ⬦, Shujun Chen *, Junbiao Shi, Hongyu Tian and Yun Zhao

College of Mechanical Engineering and Applied Electronics Technology, Beijing University of Technology, Beijing 100124, China; lif@bjut.edu.cn (F.L.); shibeard@emails.bjut.edu.cn (J.S.); jdthongyu@buu.edu.cn (H.T.); bj_ycw@emails.bjut.edu.cn (Y.Z.)
* Correspondence: sjchen@bjut.edu.cn; Tel.: +86-010-6739-1620

Received: 30 October 2017; Accepted: 23 November 2017; Published: 28 November 2017

**Featured Application: This paper proposes a hybrid manufacturing process combining wire arc additive manufacturing with milling, which provides a cost-effective and efficient way to fabricate stiffened panels that have wide applications in aviation, aerospace, and automotive industries, etc.**

**Abstract:** This paper proposes a hybrid WAAM (wire arc additive manufacturing) and milling process (HWMP), and highlights its application in the fabrication of stiffened panels that have wide applications in aviation, aerospace, and automotive industries, etc. due to their light weight and strong load-bearing capability. In contrast to existing joining or machining methods, HWMP only deposits stiffeners layer-by-layer onto an existing thin plate, followed by minor milling of the irregular surfaces, which provides the possibility to significantly improve material utilization and efficiency without any loss of surface quality. In this paper, the key performances of HWMP in terms of surface quality, material utilization and efficiency are evaluated systematically, which are the results of the comprehensive effects of the deposition parameters (e.g., travel speed, wire-feed rate) and the milling parameters (e.g., spindle speed, tool-feed rate). In order to maximize its performances, the optimization is also performed to find the best combination of the deposition and the milling parameters. The case study shows that HWMP with the optimal process parameters improves the material utilization by 57% and the efficiency by 32% compared against the traditional machining method. Thus, HWMP is believed to be a more environmental friendly and sustainable method for the fabrication of stiffened panels or other similar structures.

**Keywords:** additive manufacturing; wire arc additive manufacturing; 3D printing; hybrid manufacturing; milling; stiffened panel

---

## 1. Introduction

Additive manufacturing (AM), also known as 3D printing, offers significant advantages in terms of reduced buy-to-fly ratios, improved design flexibility, and shortened supply cycle compared to traditional subtractive manufacturing [1]. A variety of materials are now available for AM applications including metal, ceramic, plastic, etc. [2]. For metal materials, the related AM techniques mainly include selective laser sintering (SLS), selective laser melting (SLM), laser engineered net shaping (LENS), electron beam melting (EBM), electron beam freeform fabrication (EBF3), wire arc additive manufacturing (WAAM), etc. [3,4]. These techniques differ in terms of energy source (laser, electron beam, or welding arc) and feedstock (powder or wire). Nevertheless, no matter which energy source and which feedstock are adopted, it is still difficult to fabricate parts with the same level of geometric accuracy and surface quality as traditional subtractive manufacturing due to the stair-stepping effect and the liquidity of molten metal [5].

The emergence of hybrid manufacturing, integrating additive and subtractive processes into a single setup, has provided a fundamental solution to overcome the above obstacle [6–8]. Hybrid manufacturing is realized by alternating additive and subtractive processes every one or several layers, the former producing near-net-shape raw part, whereas the latter refining the raw part to achieve the desired geometric accuracy and surface quality. Hybrid manufacturing makes full use of the advantages of each individual process while minimizing their disadvantages. In recent years, various hybrid manufacturing techniques, such as hybrid layered manufacturing [9], hybrid plasma deposition and milling [10], 3D welding and milling [11], iAtractive [12], etc., have been developed. Parts with high buy-to-fly ratios or with internal and overhanging features that are difficult/expensive to fabricate with traditional manufacturing techniques will favor hybrid manufacturing.

A hybrid WAAM and milling process (HWMP, for short) is focused on in this paper. WAAM, employing welding arc as the energy source and metal wire as the feedstock for additive manufacturing purposes, is especially well known for its high productivity, low cost, high material utilization, and high energy efficiency [13,14]. Particularly, the deposition rate of WAAM can reach up to 50–130 g/min with almost no limitation of the build volume, compared to 2–10 g/min in laser- or electron beam-based processes [15]. Thus, WAAM is considered as a more economic and efficient option for fabricating medium to large-scale metal parts compared to other metal AM techniques. In recent years, WAAM has drawn significant interests from both academia and industry covering various types of materials. Cong studied the relationship between depositing mode and porosity, microstructure, and properties in WAAM of Al-6.3%Cu alloy [16]. Wu investigated the effects of heat accumulation on stability of deposition, oxidation, geometrical shape, arc characteristics, and metal transfer behavior in WAAM of Ti6Al4V alloy [17]. Xu studied the oxide accumulation mechanisms and the influence of oxides on the subsequent deposition in WAAM of maraging steel wall structure [18]. BAE systems have applied this technique to build large components, such as 1.2 m Ti6Al4V wing spar [19].

A typical application of HWMP is to fabricate stiffened panels, which have wide applications in aviation, aerospace, and automotive industries, etc., due to the advantages of light-weight and strong load-bearing capability, as shown in Figure 1a,b [20]. Generally, a stiffened panel can be fabricated either by joining the stiffeners to a thin plate via fasteners (rivets or bolts, see Figure 1c) and welding (friction stir welding or laser beam welding, see Figure 1d), or by machining from a thick plate (see Figure 1e). The joining methods have limits in reducing the total weight due to the existence of fasteners and extra flanges, whereas the machining method suffers from extremely high buy-to-fly ratios because the majority of the raw material has to be removed. In contrast to these existing methods, HWMP only deposits stiffeners layer-by-layer onto an existing thin plate followed by minor milling of the irregular surfaces, as shown in Figure 1f. This provides the possibility to significantly improve material utilization and efficiency without any loss of surface quality.

The key performance indicators of HWMP mainly include surface quality, material utilization and efficiency. Surface quality greatly affects the functional attributes of the products including friction, wear resistance, fatigue, lubricant, light reflection and coating [21]. Material utilization and efficiency are strongly related to the environmental and economic benefits [22]. The challenge for evaluating these indicators lies in that the deposition parameters (e.g., travel speed, wire-feed rate) and the milling parameters (e.g., spindle speed, tool-feed rate) have comprehensive effects on them. For example, the surface quality achieved by HWMP is dependent upon not only spindle speed and tool-feed rate, just like other independent milling processes, but also travel speed and wire-feed rate. This is because the cutting depth in the milling step is directly determined by the bead geometry produced in the previous deposition step, which is a function of travel speed and wire-feed rate.

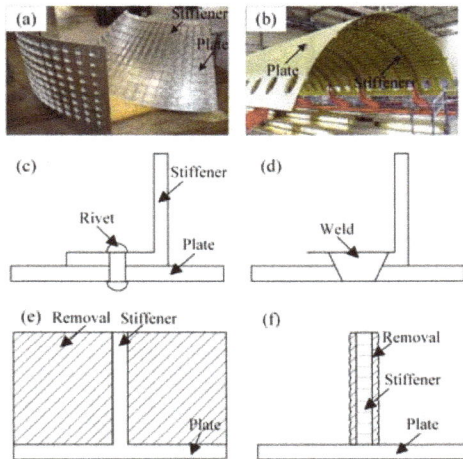

**Figure 1.** Processing technologies for stiffened panels. (**a,b**) Applications of stiffened panels [23]; (**c**) riveting; (**d**) welding; (**e**) machining; and (**f**) HWMP.

The primary aim of this paper is to evaluate the efficacy of HWMP in the fabrication of stiffened panels. The surface quality, material utilization and efficiency are evaluated systematically with consideration of the comprehensive effects of the deposition and the milling parameters. In addition, the optimization is also performed based on a genetic algorithm (GA) to find the best combination of the deposition and the milling parameters in order to maximize its performances.

## 2. System Description

Figure 2 shows a two-robot cooperative platform for implementing HWMP, developed at Beijing University of Technology (BJUT). The welding torch and the milling tool are mounted on two independent robots. The first robot is RTI2000 (igm Robotersysteme AG, Wiener Neudorf, Austria), equipped with two Fronius Synergic 5000 power sources to implement WAAM based on Tandem GMAW (gas metal arc welding). Tandem GMAW differs from conventional GMAW as two welding wires are passed through the same welding torch and, therefore, provides much higher productivity [24]. Based on preliminary experiments, it is known that tandem GMAW is capable to produce wall structures of widths ranging from about 4 mm to 17 mm, which is especially beneficial for fabricating stiffened panels of different specifications. The wire material used in this study is Al2325 alloy with the chemical composition of Cu 3.9–4.8%, Mn 0.1–1.0%, Ti 0.15%, Mg 0.4–0.8%, Zn 0.3%, etc., in addition to Al, and the substrate material is Al2219 alloy. The wire diameter is 1.2 mm, and the shielding gas is Ar at a rate of 22 L/min. Aluminum alloys are one of the most widely used materials in aircraft components because of their reasonable cost, high stiffness-to-weight and strength-to-weight ratios, and excellent machinability. The second robot is KR500 (KUKA AG, Augsburg, Bavaria, Germany), which is a heavy-duty robot that is suitable for milling applications. It is equipped with a high-speed electric spindle ES779 with a maximum spindle speed of 22,000 rpm. The uncoated carbide alloy milling tool is adopted, which has a diameter of 14 mm and a helix angle of 55°. The working mode is down milling. No cooling and lubricating agent are used.

The work principle of HWMP for fabricating stiffened panels is displayed in Figure 3. Step 1: the welding torch moves along the length of the stiffeners and deposits *N* layers onto an existing thin plate or the previous layers; Step 2: the top surface of the deposited layers is milled to a prescribed thickness *H* in order to facilitate the subsequent deposition; Step 3: the two side surfaces of the deposited layers are milled to obtain the desired geometric accuracy and surface quality. These steps alternate until the whole part is created. It should be emphasized that the deposited part must cool down to room temperature before next deposition or milling to avoid excessive heat accumulation. If $N$ is too small, the alternation of deposition and milling would be repeated many times, which decreases the productivity; but if too large, the axial cutting depth is also too large, which increases the cutting force. Therefore, $N$ is determined to be six in this study. The total thickness of six layers is about 12 mm (2 mm for one layer). $H$ is determined to be 8 mm such that the irregular surfaces could be removed completely.

**Figure 2.** Two-robot cooperative platform for implementing HWMP.

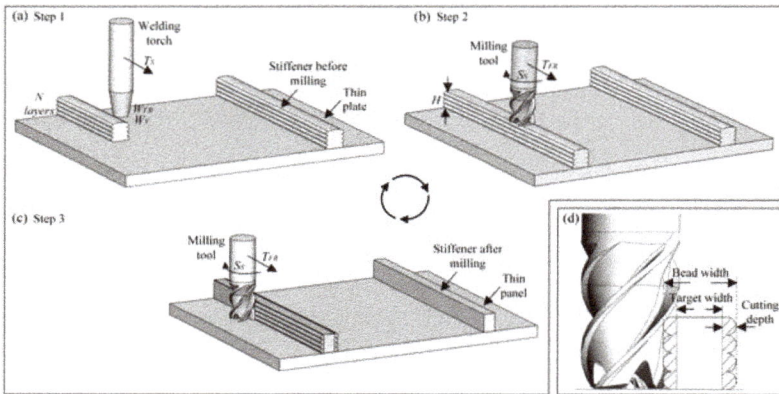

**Figure 3.** (**a–c**) Work principle of HWMP for fabricating stiffened panels; and (**d**) the relation between cutting depth and bead width.

## 3. Evaluation of Surface Quality, Material Utilization, and Efficiency

### 3.1. Evaluation of Surface Quality

Generally, the key deposition parameters affecting the bead geometry in Step 1 mainly include wire-feed rate ($W_{FR}$), travel speed ($T_S$), and welding voltage ($W_V$). The key milling parameters affecting the surface quality in Step 2 and Step 3 mainly include spindle speed ($S_S$), tool-feed rate ($T_{FR}$), and cutting depth ($C_D$). Only the side surface's quality in Step 3 is concerned in this study because the top surface will be covered by subsequent layers. Unlike other independent milling processes, the cutting depth in Step 3 is directly determined by the bead geometry produced in Step 1, as shown in Figure 3d. The larger the bead width ($B_W$) than the target width ($T_W$), the larger the cutting depth. The bead width is determined by the deposition parameters and the target width is a constant value for a specific stiffened panel. Therefore, the surface quality (represented by surface roughness, $R_a$) achieved by HWMP is a result of both the deposition and the milling parameters, as seen in Figure 4.

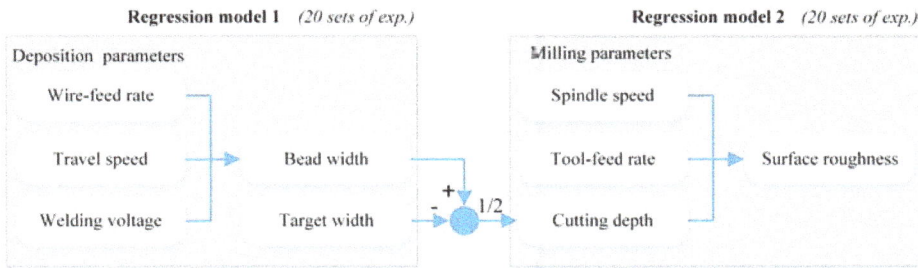

**Figure 4.** Relation between surface roughness and the key process parameters.

To model the surface roughness in an efficient way, two cascaded regression models are developed, the first model relating the bead width to the deposition parameters and the second one relating the surface roughness to the milling parameters, as seen in Figure 4. The output of the first model, i.e., bead width, determines the input of the second model, i.e., cutting depth. Then the two models are synthesized to obtain the surface roughness model by establishing a link between cutting depth and bead width. The central composite rotatable design (CCRD) method is applied to obtain each regression model, which has been demonstrated to be an efficient experiment design method with a relatively small number of experiments without losing its accuracy [25,26]. To apply CCRD, the following procedure should be obeyed: (1) identifying predominant factors affecting the response; (2) determining their upper and lower limits; (3) generating experimental design matrix; (4) conducting experiments according to the experimental design matrix; (5) developing the regression model; and (6) validating the adequacy of the developed model by analysis of variance (ANOVA).

3.1.1. Identifying Predominant Factors Affecting the Response and Determining Their Limits

The predominant factors affecting the response in each regression model have been discussed above. Their upper and lower limits that are of interest in this study are given in Table 1. These factors are coded as −1.68, −1, 0, +1, and +1.68. Specifically, the ranges of the three deposition parameters are determined based on preliminary experiments, which allow good bead appearance with little spatter and no visible defects. It should be pointed out the bead width is not a constant value along the build direction due to the presence of the stair-stepping effect, as shown in Figure 3d. To address this issue, the average bead width is used, which is defined as the ratio of the cross-section area to the height of the produced wall.

**Table 1.** Coding for factor and level.

| Symbol | Factor | Unit | Level | | | | |
|--------|--------|------|-------|---|---|---|---|
| | | | −1.68 | −1 | 0 | 1 | 1.68 |
| **Regression model 1 (Response: bead width)** | | | | | | | |
| $W_{FR}$ | Wire-feed rate | m/min | 3.4 | 3.7 | 4.3 | 4.8 | 5.1 |
| $T_S$ | Travel speed | m/min | 0.35 | 0.4 | 0.48 | 0.55 | 0.6 |
| $W_V$ | Welding voltage | V | 16.6 | 17.3 | 18.4 | 19.5 | 20.2 |
| **Regression model 2 (Response: surface roughness)** | | | | | | | |
| $S_S$ | Spindle speed | rpm | 1000 | 2400 | 4500 | 6600 | 8000 |
| $T_{FR}$ | Tool-feed rate | mm/s | 1 | 1.8 | 3 | 4.2 | 5 |
| $C_D$ | Cutting depth | mm | 1 | 1.4 | 2 | 2.6 | 3 |

3.1.2. Generating Experimental Design Matrix and Conducting the Experiments

For three factors chosen with five levels, the required number of experiments is 20 according to CCRD, eight as factorial points, six as star points, and six as center points. As a consequence, the two cascaded regression models require 20 sets of deposition experiments and 20 sets of milling experiments in total. The resulting experimental design matrix is given in Table 2.

**Table 2.** Experimental design matrix and the response.

| | Regression Model 1 | | | Regression Model 2 | |
|---|---|---|---|---|---|
| Exp. No. | Coding ($W_{FR}\ T_S\ W_V$) | Bead Width (mm) | Exp. No. | Coding ($S_S\ T_{FR}\ C_D$) | Roughness (μm) |
| 1 | (−1 −1 −1) | 9 | 1 | (−1 −1 −1) | 1.74 |
| 2 | (1 −1 −1) | 11.9 | 2 | (1 −1 −1) | 1.41 |
| 3 | (−1 1 −1) | 8.4 | 3 | (−1 1 −1) | 1.99 |
| 4 | (1 1 −1) | 10.8 | 4 | (1 1 −1) | 1.47 |
| 5 | (−1 −1 1) | 9.5 | 5 | (−1 −1 1) | 1.97 |
| 6 | (1 −1 1) | 11.7 | 6 | (1 −1 1) | 1.58 |
| 7 | (−1 1 1) | 8.3 | 7 | (−1 1 1) | 2.15 |
| 8 | (1 1 1) | 10 | 8 | (1 1 1) | 1.81 |
| 9 | (−1.682 0 0) | 8 | 9 | (−1.682 0 0) | 2.41 |
| 10 | (1.682 0 0) | 12.5 | 10 | (1.682 0 0) | 1.52 |
| 11 | (0 −1.682 0) | 11.5 | 11 | (0 −1.682 0) | 1.79 |
| 12 | (0 1.682 0) | 9.5 | 12 | (0 1.682 0) | 1.86 |
| 13 | (0 0 −1.682) | 9.9 | 13 | (0 0 −1.682) | 1.68 |
| 14 | (0 0 1.682) | 10.1 | 14 | (0 0 1.682) | 1.78 |
| 15 | (0 0 0) | 9.6 | 15 | (0 0 0) | 1.65 |
| 16 | (0 0 0) | 9.5 | 16 | (0 0 0) | 1.64 |
| 17 | (0 0 0) | 9.6 | 17 | (0 0 0) | 1.67 |
| 18 | (0 0 0) | 10 | 18 | (0 0 0) | 1.53 |
| 19 | (0 0 0) | 9.5 | 19 | (0 0 0) | 1.52 |
| 20 | (0 0 0) | 10.1 | 20 | (0 0 0) | 1.59 |

According to Column 1–2 in Table 2, 20 sets of deposition experiments were carried out first. A total of six layers were deposited in each experiment, as seen in Figure 5a. The bead width along the build direction was measured with the aid of a laser displacement scanner (HG-C1030, Panasonic, 0.01 mm repeatable precision) and then the average value was calculated, as given in Column 3 of Table 2.

With the wall structures produced by WAAM, 20 sets of milling experiments were carried out subsequently according to Column 4–5 in Table 2, as seen in Figure 5b. The surface roughness in the tool-feed direction was measured with a roughmeter (TR200, 0.01 μm sensitivity, Time, Beijing, China). To eliminate random errors, the surface roughness was measured five times at different locations and repeated twice at each location. The results are given in Column 6 of Table 2.

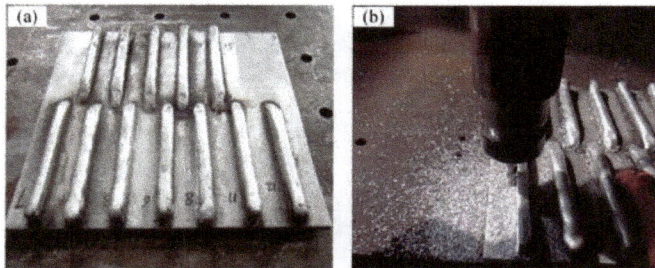

**Figure 5.** (**a**) Wall structures produced by WAAM; and (**b**) milling experiments.

### 3.1.3. Developing and Validating the Regression Models

Based on the results of the measured bead width given in Table 2, the regression model (second-order) that describes the dependence of $B_W$ on $W_{FR}$, $T_S$ and $W_V$ is obtained with the aid of the software Design-Expert (Version 6.0, State-Ease, Minneapolis, MN, USA, 2005) as follows:

$$B_W = 9.73 + 1.23W_{FR} - 0.58T_S - 0.019W_V - 0.12W_{FR}T_S - 0.17W_{FR}W_V - 0.15T_SW_V \\ + 0.12W_{FR}^2 + 0.21T_S^2 + 0.03W_V^2 \tag{1}$$

Then ANOVA is undertaken for validating this model and the results are given in Column 1–3 of Table 3. It is seen that the $F$-value of the model is 33.54, much higher than $F_{0.05}(9, 10) = 3.179$, indicating that this model is significant at a 95% confidence level, whereas $F$-value of lack of fit is 1.52, lower than $F_{0.05}(5, 5) = 5.05$, indicating that lack of fit is not significant. Moreover, the coefficient of determination $R^2$ is very close to 1, i.e., $R^2 = 0.9679$, which means that the model clarifies 96.79% of all deviations. Thus, we can conclude that this obtained regression model is credible and accurate. At a 95% confidence level, only $p$-values of $T_S$, $W_{FR}$, and $T_S^2$ term are all lower than 0.05, which indicate that only their effects on $B_W$ are significant. After omitting the insignificant terms, this regression model is simplified to:

$$B_W = 9.73 + 1.23W_{FR} - 0.58T_S + 0.21T_S^2 \tag{2}$$

**Table 3.** ANOVA results of the two regression models.

| Regression Model 1 | | | Regression Model 2 | | |
|---|---|---|---|---|---|
| Source | $F$-Value | $p$-Value | Source | $F$-Value | $p$-Value |
| $A$-$W_{FR}$ | 234.16 | <0.0001 | $A$-$S_S$ | 66.73 | <0.0001 |
| $B$-$T_S$ | 52.82 | <0.0001 | $B$-$T_{FR}$ | 5.11 | 0.0473 |
| $C$-$W_V$ | 0.058 | 0.8147 | $C$-$C_D$ | 8.20 | 0.0169 |
| $AB$ | 1.42 | 0.2607 | $AB$ | 0.23 | 0.6390 |
| $AC$ | 2.79 | 0.1260 | $AC$ | 0.16 | 0.7020 |
| $BC$ | 2.05 | 0.1830 | $BC$ | 0.080 | 0.7827 |
| $A^2$ | 2.30 | 0.1604 | $A^2$ | 16.72 | 0.0022 |
| $B^2$ | 7.01 | 0.0244 | $B^2$ | 5.13 | 0.0469 |
| $C^2$ | 0.15 | 0.7086 | $C^2$ | 1.04 | 0.3318 |
| **Model** | 33.54 | <0.0001 | **Model** | 11.21 | 0.0004 |
| **Lack of Fit** | 1.52 | 0.3275 | **Lack of Fit** | 4.33 | 0.0669 |
| $R^2$ | 0.9679 | | $R^2$ | 0.9099 | |

Analogously, the regression model (second-order) that describes the dependence of $R_a$ on $S_S$, $T_{FR}$ and $C_D$ is also obtained as follows:

$$R_a = 1.60 - 0.23S_S + 0.063T_{FR} + 0.079C_D - 0.017S_ST_{FR} + 0.014S_SC_D + 0.010T_{FR}C_D \\ + 0.11S_S^2 + 0.061T_{FR}^2 + 0.027C_D^2 \tag{3}$$

The corresponding ANOVA results are given in Column 4–6 of Table 3, which prove that this regression model is also credible and accurate. Only $S_S$, $T_{FR}$, $C_D$, $S_S^2$ and $T_{FR}^2$ have significant effects on $R_a$ at 95% confidence level and as a result the simplified regression model is

$$R_a = 1.60 - 0.23S_S + 0.063T_{FR} + 0.079C_D + 0.11S_S^2 + 0.061T_{FR}^2 \tag{4}$$

### 3.1.4. Developing Surface Roughness Model

From Figure 3d, it is clearly seen that the bead width and the cutting depth satisfy the following relation:

$$C_D = (B_W - T_W)/2 \tag{5}$$

Through replacing $C_D$ in Equation (4) with Equation (5) and combining Equation (2), the final surface roughness model is obtained:

$$R_a = 1.98 - 0.065T_W + 0.08W_{FR} - 0.0377T_S + 0.0137T_S^2 - 0.23S_S + 0.11S_S^2 + 0.063T_{FR} + 0.061T_{FR}^2 \quad (6)$$

which clearly exhibits the dependence of the surface roughness on both the deposition and the milling parameters. In addition, the target width also affects the surface roughness.

### 3.2. Evaluation of Material Utilization

Material utilization ($M_U$) is defined as the ratio of the final part's mass ($m_{part}$) to the raw material's mass ($m_{raw\_material}$) as follows:

$$M_U = \frac{m_{part}}{m_{raw\_material}} \quad (7)$$

The final part's mass is the sum of the masses of the plate ($m_{plate}$) and the stiffeners ($m_{stiffener}$), whereas the raw material's mass is the sum of the masses of the plate and the beads ($m_{bead}$). From Figure 3d, we also know that $m_{bead}/m_{stiffener}$ is approximately equal to $B_W/T_W$, neglecting the removed mass in Step 2. Therefore, Equation (7) can be converted to:

$$
\begin{aligned}
M_U &= \frac{m_{plate} + m_{stiffener}}{m_{plate} + m_{bead}} \approx \frac{m_{plate} + m_{stiffener}}{m_{plate} + \frac{B_W}{T_W}m_{stiffener}} \\
&= \frac{\frac{m_{plate}}{m_{stiffener}} + 1}{\frac{m_{plate}}{m_{stiffener}} + \frac{B_W}{T_W}} = \frac{\frac{m_{plate}}{m_{stiffener}} + 1}{\frac{m_{plate}}{m_{stiffener}} + \frac{9.73 + 1.23W_{FR} - 0.58T_S + 0.21T_S^2}{T_W}}
\end{aligned}
\quad (8)
$$

which is a function of the wire-feed rate, travel speed, target width, and the ratio of the masses of the plate to the stiffeners.

### 3.3. Evaluation of Efficiency

Regarding the efficiency (represented by construction time, $T_C$), it is related to two main process parameters, i.e., travel speed and tool-feed rate, the former determining the deposition time ($T_{deposition}$), whereas the latter determining the milling time ($T_{milling}$). Additionally, the cooling time ($T_{cooling}$) for the part to cool down to room temperature before next deposition or milling and the tool-changing time ($T_{tool-changing}$) for switching the welding torch and the milling tool should also be considered. Therefore, the construction time that the deposition and the milling processes alternate once (i.e., $N = 6$) can be calculated as follows:

$$
\begin{aligned}
T_C &= T_{deposition} + T_{milling} + T_{cooling} + T_{tool\_changing} \\
&= \frac{6L}{T_S} + \frac{3L}{T_{FR}} + T_{cooling} + T_{tool\_changing}
\end{aligned}
\quad (9)
$$

where the coefficient 6 in the first term denotes that six layers are deposited, the coefficient 3 in the second term denotes that both the top surface and the two side surfaces are milled and $L$ represents the length of the stiffeners. The last two terms are assumed to remain unchanged regardless of the variation of the process parameters.

### 3.4. Effects of Process Parameters on the Performances

Based on Equations (6), (8) and (9), the effects of single process parameter on surface roughness, material utilization, and construction time are analyzed, as shown in Figure 6. This is obtained by varying one process parameter while keeping the other process parameters at zero level (according to Table 1). The other unknown parameters are set as follows as an example:

$$T_W = 7 \text{ mm}, L = 1200 \text{ mm}, \frac{m_{plate}}{m_{stiffener}} = 4, t_{cooling} = 15 \text{ min}, t_{tool\_changing} = 2 \text{ min} \quad (10)$$

From Figure 6a, it is known that the surface roughness appears to be a nonlinear decreasing function of travel speed and spindle speed. Higher travel speed leads to smaller bead width according to Equation (2) and, therefore, smaller cutting depth according to Equation (5). Both smaller cutting depth and higher spindle speed contribute to lower surface roughness according to Equation (4). On the other hand, the surface roughness increases almost linearly with increased wire-feed rate. This is because higher wire-feed rate leads to larger cutting depth according to Equations (2) and (5) and, therefore, higher surface roughness according to Equation (4). Additionally, the surface roughness decreases slightly and then increases greatly with the increased tool-feed rate. These results basically agree with the observations in previous research. A reasonable increase of spindle speed would decrease the cutting force due to the thermal softening and, therefore, lead to lower surface roughness [27]. An increase of cutting depth and tool-feed rate means removing more volume per unit time, which would result in larger cutting force and, therefore, higher surface roughness [28]. From Figure 6, we can also know that the steeper the slope of a curve, the larger the corresponding parameter's contribution to the surface roughness. To make a fair comparison, the slopes of these curves are compared at zero level. Then it can be obtained that the order of these parameters' contribution to the surface roughness is $S_S > W_{FR} > T_{FR} > T_S$.

Figure 6b shows the dependence of the material utilization on travel speed and wire-feed rate. Higher wire-feed rate and lower travel speed are likely to lead to larger bead width according to Equation (2) and, therefore, lower material utilization. The two process parameters must match each other in order to ensure a high material utilization. By comparing the slopes of the two curves, it is known that the wire-feed rate plays a dominant role in determining the material utilization. From Figure 6c, it is observed that the construction time is a decreasing function of both the travel speed and the tool-feed rate, which is easy to explain by combining Equation (8). Additionally, the tool-feed rate plays a dominant role in determining the construction time.

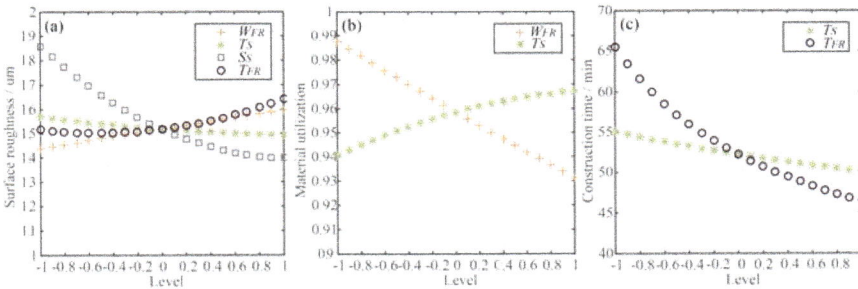

**Figure 6.** (**a**) The effects of single process parameter on surface roughness; (**b**) the effects of single process parameter on material utilization; and (**c**) the effects of single process parameter on construction time.

## 4. Parameter Optimization

The optimization is performed to find best combination of the deposition and the milling parameters in order to maximize the performances of HWMP in terms of surface quality, material utilization, and efficiency. The optimization problem can be expressed in the following form:

Objective function:

$$\min R_a(W_{FR}, T_S, S_S, T_{FR}), \min 1/M_U(W_{FR}, T_S) \text{ and } \min T_C(T_S, T_{FR}) \tag{11}$$

within the ranges of these process parameters:

$$3.4 \text{ m/min} \le W_{FR} \le 5.1 \text{ m/min}$$

$$0.35 \, \text{m/min} \leq T_S \leq 0.6 \, \text{m/min}$$

$$1000 \, \text{rpm} \leq S_S \leq 8000 \, \text{rpm}$$

$$1 \, \text{mm/s} \leq T_{FR} \leq 5 \, \text{mm/s}$$

and subjected to the constraint function:

$$T_W + 1 \leq 9.73 - 0.58 T_S + 1.23 W_{FR} + 0.21 T_S^2 \, (i.e., B_W) \tag{12}$$

The constraint function (Equation (12)) means that the achieved bead width should be at least 1 mm larger than the target width to ensure that the "stair-stepping effect" can be removed completely.

GA is adopted to solve this optimization problem that is based on natural selection and biological evolution [29,30]. GA has become popular in engineering optimization problems for its wide range of precise search and capability of solving complex non-linear problems. Due to the complexity of the multi-objective optimization problem, the weighted sum method (WSM) is applied first to convert the three objectives (i.e., min $R_a$, min $1/M_U$, and min $T_C$) to a single objective (i.e., min ($w_1 R_a + w_2 / M_U + w_3 T_C$)) [31]. $w_1$, $w_2$, and $w_3$ are the weight coefficients and $w_1 + w_2 + w_3 = 1$. These parameters should be normalized before being converted as follows:

$$R_a^* = \frac{R_a - R_{a\_min}}{R_{a\_max} - R_{a\_min}} \tag{13}$$

$$M_U^* = \frac{M_U - M_{U\_min}}{M_{U\_max} - M_{U\_min}} \tag{14}$$

$$T_C^* = \frac{T_C - T_{C\_min}}{T_{C\_max} - T_{C\_min}} \tag{15}$$

Then the resultant objective function is obtained as follows:

$$\begin{aligned} &\min \left( w_1 R_a^* + w_2 / M_U^* + w_3 T_C^* \right) \\ &= \min \left( w_1 \frac{R_a - R_{a\_min}}{R_{a\_max} - R_{a\_min}} + w_2 / \frac{M_U - M_{U\_min}}{M_{U\_max} - M_{U\_min}} + w_3 \frac{T_C - T_{C\_min}}{T_{C\_max} - T_{C\_min}} \right) \end{aligned} \tag{16}$$

where $w_1$ is set to 0.6, $w_2$ is set to 0.2 and $w_3$ is set to 0.2 in this study.

By using the MATLAB optimization toolbox, the optimization solutions for different target widths are obtained as given in Table 4. The parameters such as $L$, $m_{\text{plate}}/m_{\text{stiffener}}$, etc., are the same as those in Equation (10). From Table 4, we can observe that for any target width, a high travel speed, and a high spindle speed are recommended because the surface roughness is a decreasing function of the two process parameters according to Figure 6a. Higher travel speed also results in both higher material utilization and efficiency according to Figure 6b,c. Though a low tool-feed rate is beneficial for achieving a low surface roughness according to Figure 6a, it also increases the construction time according to Figure 6c. Therefore, a moderate tool-feed rate is recommended to obtain a good balance between surface roughness and construction time. With regard to wire-feed rate, different target widths correspond to different wire-feed rates, which can be explained through the constraint function (Equation (12)), i.e., larger target width required larger bead width and therefore larger wire-feed rate. In all, a high travel speed, a target width dependent wire-feed rate, a high spindle speed and a moderate tool-feed rate are required to maximize the performances of HWMP. It is interesting to find that the optimized values for the surface roughness, the material utilization and the construction time are basically the same regardless of the target width. The surface roughness remains around 1.3 µm and the material utilization ranges from 96% to 98%.

**Table 4.** Optimization results under different target widths.

| $T_W$ (mm) | $W_{FR}$ (m/min) | $T_S$ (m/min) | $S_S$ (rpm) | $T_{FR}$ (mm/s) | $R_a$ (μm) | $M_U$ | $T_C$ (min) |
|---|---|---|---|---|---|---|---|
| 6 | 3.5 | 0.6 | 6694 | 3.1 | 1.31 | 96% | 48.35 |
| 7 | 3.8 | 0.6 | 6694 | 3.1 | 1.29 | 97% | 48.35 |
| 8 | 4.2 | 0.6 | 6695 | 3.1 | 1.29 | 97% | 48.35 |
| 9 | 4.6 | 0.6 | 6695 | 3.1 | 1.29 | 98% | 48.35 |
| 10 | 5.0 | 0.6 | 6695 | 3.1 | 1.29 | 98% | 48.35 |

## 5. Case Study

For validation of the developed models and the optimization results above, a test part with intersecting stiffeners was fabricated following the steps in Figure 3, as shown in Figure 7. Note that the width and the total length of the stiffeners are 7 mm and 1200 mm respectively, and the ratio of the masses of the plate to the stiffeners is about 4, which are the same as those in Equation (10). Therefore, the corresponding optimal process parameters are $W_{FR}$ = 3.8 m/min, $T_S$ = 0.6 m/min, $S_S$ = 6694 rpm and $T_{FR}$ = 3.1 mm/s according to Table 4. Furthermore, note that the height of each stiffener is 16 mm, which means that the deposition and the milling processes need to alternate twice here.

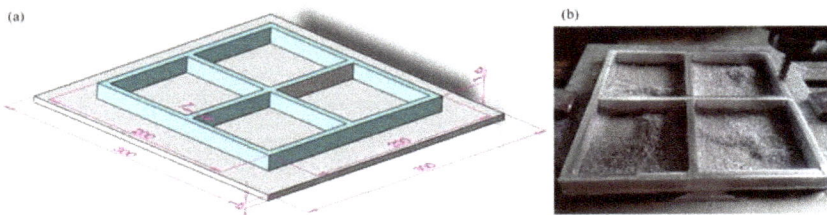

**Figure 7.** (a) Stereogram; (b) photograph of the produced stiffened panel.

After this stiffened panel was fabricated, the actual surface roughness in the tool-feed direction was measured at different locations. The average value is 1.38 μm, which is about 7.0% higher than the predicted value, i.e., 1.29 μm. The actual material utilization is calculated by dividing the final part's mass (1.82 kg) by the sum of the masses of the plate and the raw metal wire (1.46 kg + 0.55 kg). The result is 91%, which is a little lower than the predicted value, i.e., 97%. This is attributed to the fact that the removed mass from the top surface in Step 2 is neglected in the theoretical evaluation. The actual construction time is about 102 min, which is slightly longer than the predicted value (48.35 × 2 = 96.7 min). This is because the time for the milling tool to move from one surface to another is not considered in the theoretical evaluation. In all, these developed models (Equations (6), (8) and (9)) are capable of predicting the surface quality, material utilization, and efficiency of HWMP with reasonable accuracy.

As a comparison, the performances of the traditional machining for the same stiffened panel are also evaluated. It starts with a thick plate with dimensions 300 mm × 300 mm × 22 mm and then goes through two stages: roughing and finishing. The roughing stage aims to remove extra mass as quickly as possible by employing a large material removal rate (MRR), whereas the finishing stage aims to refine the part to the required accuracy by employing a small MRR. To make a fair comparison, the MRR selected for the finishing stage is the same as that in HWMP, i.e., $MRR_{finishing}$ = 14.88 mm³/s. That is to say, the achieved surface quality here is assumed to be the same as that in HWMP. The MRR for the roughing stage, i.e., $MRR_{roughing}$, is assumed to be 10 times that of $MRR_{finishing}$, i.e., 148.8 mm³/s. The machining time for each stage can be calculated by dividing the volume to be removed by the corresponding MRR. The tool-changing time is not considered here.

A detailed comparison between HWMP and the traditional machining is summarized in Table 5. HMWP improves the material utilization from 34% to 91% (by 57%) and reduces the construction time from 166 min to 102 min (by 32%). Significant advantages of HWMP over the traditional machining for the fabrication of stiffened panels are demonstrated. It is also observed that the cooling time accounts for about one third of the total construction time in HWMP. To further increase its efficiency, it is necessary to introduce additional cooling methods, such as forced air cooling, to reduce the cooling time as much as possible.

**Table 5.** Comparison between HWMP and the traditional machining for the stiffened panel.

| Traditional Machining | | HWMP | |
|---|---|---|---|
| Thick plate's mass | 5.35 kg | Thin plate's mass | 1.46 kg |
| | | Metal wire's mass | 0.55 kg |
| Final part's mass | 1.82 kg | Final part's mass | 1.82 kg |
| Material utilization | 34% | Material utilization | 91% |
| | | Deposition time | 24 min |
| Roughing time | 144 min | Milling time | 38.7 min |
| Finishing time | 22 min | | |
| | | Cooling, etc. | 39.3 min |
| Construction time | 166 min | Construction time | 102 min |

## 6. Conclusions and Future Work

A hybrid manufacturing process combining WAAM and milling (HWMP) is proposed in this paper, which provides an entirely new method to fabricate stiffened panels in contrast to existing joining or machining methods. The comprehensive effects of the deposition parameters (including wire-feed rate and travel speed) and the milling parameters (including spindle speed and tool-feed rate) on surface quality, material utilization, and efficiency are investigated systematically. Only good matching of the deposition and the milling parameters could maximize the performances of HMWP. Through the optimization, it is obtained that a high travel speed, a target width dependent wire-feed rate, a high spindle speed, and a moderate tool-feed rate are required to obtain a good balance between surface quality, material utilization, and efficiency. This provides a guide to select approximate process parameters for hybrid manufacturing processes. With the optimal process parameters, significant reduction in material consumption and improvement in efficiency can be realized without any loss of accuracy for the fabrication of stiffened panels, compared against the traditional machining. Therefore, HWMP that accords with the concept of sustainable development would be attractive to the manufacturers.

Despite the above advantages, the WAAM process based on welding may introduce undesirable residual stresses and, therefore, distortions due to the large heat input, especially for thin structures. Several measures could be taken to solve this problem. For example, some new welding techniques with lower heat input could be used instead of the conventional GMAW, such as cold metal transfer (CMT), dual-bypass GMAW and hot wire GMAW [32]. Some preheating approaches such as induction heating could also be used to regulate the thermal gradient. Additionally, well-designed fixtures are necessary to reduce the distortions. These works will be considered in the future.

**Acknowledgments:** This paper was supported by the National Natural Science Foundation of China (No. 51475009) and a China Postdoctoral Science Foundation funded project (No. 2017M610726).

**Author Contributions:** Fang Li performed the modeling and wrote the manuscript; Shujun Chen contributed to the optimization and the analysis of the data; Junbiao Shi designed and performed the deposition experiments; Hongyu Tian designed and performed the milling experiments; and Yun Zhao helped perform the experiments and analyze the data.

**Conflicts of Interest:** The authors declare no conflict of interest.

## References

1. Gao, W.; Zhang, Y.; Ramanujan, D.; Ramani, K.; Williams, C.B. The status, challenges, and future of additive manufacturing in engineering. *Comput. Aided Des.* **2015**, *69*, 65–89. [CrossRef]
2. Singh, S.; Ramakrishna, S.; Singh, R. Material issues in additive manufacturing: A review. *J. Manuf. Process.* **2017**, *25*, 185–200. [CrossRef]
3. Frazier, W.E. Metal Additive Manufacturing: A Review. *J. Mater. Eng. Perform.* **2014**, *23*, 1917–1928. [CrossRef]
4. Ding, D.; Pan, Z.; Cuiuri, D.; Li, H. Wire-feed additive manufacturing of metal components: Technologies, developments and future interests. *Int. J. Adv. Manuf. Technol.* **2015**, *81*, 465–481. [CrossRef]
5. Kumbhar, N.N.; Mulay, A.V. Post processing methods used to improve surface finish of products which are manufactured by additive manufacturing technologies: A review. *J. Inst. Eng.* **2016**, 1–7. [CrossRef]
6. Flynn, J.M.; Shokrani, A.; Newman, S.T.; Dhokia, V. Hybrid additive and subtractive machine tools—Research and industrial developments. *Int. J. Mach. Tools Manuf.* **2016**, *101*, 79–101. [CrossRef]
7. Zhu, Z.; Dhokia, V.; Nassehi, A.; Newman, S.T. A review of hybrid manufacturing processes-state of the art and future perspectives. *Int. J. Comput. Integr. Manuf.* **2013** *26*, 596–615. [CrossRef]
8. Manogharan, G.; Wysk, R.A.; Harrysson, O.L.A. Additive manufacturing–integrated hybrid manufacturing and subtractive processes: Economic model and analysis. *Int. J. Comput. Integr. Manuf.* **2016**, *29*, 473–488. [CrossRef]
9. Kapil, S.; Legesse, F.; Kulkarni, P.; Joshi, P.; Desai, A.; Karunakaran, K.P. Hybrid-layered manufacturing using tungsten inert gas cladding. *Prog. Addit. Manuf.* **2016**, *1*, 79–91. [CrossRef]
10. Xiong, X.; Zhang, H.; Wang, G.; Wang, G. Hybrid plasma deposition and milling for an aeroengine double helix integral impeller made of superalloy. *Robot. Comput. Integr. Manuf.* **2010**, *26*, 291–295. [CrossRef]
11. Song, Y.; Park, S.; Choi, D.; Jee, H. 3D welding and milling: Part I—A direct approach for freeform fabrication of metallic prototypes. *Int. J. Mach. Tools Manuf.* **2005**, *45*, 1057–1062. [CrossRef]
12. Zhu, Z.; Dhokia, V.; Newman, S.T.; Nassehi, A. Application of a hybrid process for high precision manufacture of difficult to machine prismatic parts. *Int. J. Adv. Manuf. Technol.* **2014**, *74*, 1115–1132. [CrossRef]
13. Pan, Z.; Ding, D.; Wu, B.; Cuiuri, D.; Li, H.; Norrish, J. Arc Welding Processes for Additive Manufacturing: A Review. In *Transactions on Intelligent Welding Manufacturing*; Chen, S., Zhang, Y., Feng, Z., Eds.; Springer: Singapore, 2017; ISBN 978-981-10-5355-9.
14. Wu, Q.; Ma, Z.; Chen, G.; Liu, C.; Ma, D.; Ma, S. Obtaining fine microstructure and unsupported overhangs by low heat input pulse arc additive manufacturing. *J. Manuf. Process.* **2017**, *27*, 198–206. [CrossRef]
15. Ding, D.; Pan, Z.; Cuiuri, D.; Li, H. A tool-path generation strategy for wire and arc additive manufacturing. *Int. J. Adv. Manuf. Technol.* **2014**, *73*, 173–183. [CrossRef]
16. Cong, B.; Qi, Z.; Qi, B.; Sun, H.; Zhao, G.; Ding, J. A Comparative study of additively manufactured thin wall and block structure with Al-6.3%Cu alloy Using cold metal transfer process. *Appl. Sci.* **2017**, *7*, 275. [CrossRef]
17. Wu, B.; Ding, D.; Pan, Z.; Cuiuri, D.; Li, H.; Han, J.; Fei, Z. Effects of heat accumulation on the arc characteristics and metal transfer behavior in wire arc additive manufacturing of Ti6Al4V. *J. Mater. Process. Technol.* **2017**, *250*, 304–312. [CrossRef]
18. Xu, X.; Ding, J.; Ganguly, S.; Diao, C.; Williams, S. Oxide accumulation effects on wire + arc layer-by-layer additive manufacture process. *J. Mater. Process. Technol.* **2017**, *252*, 739–750. [CrossRef]
19. Williams, S.W.; Martina, F.; Addison, A.C.; Ding, J.; Pardal, G.; Colegrove, P. Wire + arc additive manufacturing. *Mater. Sci. Technol.* **2016**, *7*, 641–647. [CrossRef]
20. Moreira, P.M.G.P.; Richtertrummer, V.; Castro, P.M.S.T.C. Lightweight stiffened panels fabricated using emerging fabrication technologies: Fatigue behaviour. *Adv. Struct. Mater.* **2010**, *8*, 151–172.
21. Öktem, H. An integrated study of surface roughness for modelling and optimization of cutting parameters during end milling operation. *Int. J. Adv. Manuf. Technol.* **2009**, *43*, 852–861. [CrossRef]
22. Jin, Y.; Du, J.; He, J. Optimization of process planning for reducing material consumption in additive manufacturing. *J. Manuf. Syst.* **2017**, *44*, 65–78. [CrossRef]
23. Zhao, C.; Li, J. The manufacturing technology of integral panel on spacecraft. *Aerosp. Manuf. Technol.* **2006**, *4*, 44–48.

24. Sproesser, G.; Chang, Y.J.; Pittner, A.; Finkbeiner, M.; Rethmeier, M. Environmental energy efficiency of single wire and tandem gas metal arc welding. *Weld. World* **2017**, *61*, 733–743. [CrossRef]

25. Cukor, G.; Jurkovi, Z.; Sekuli, M. Rotatable central composite design of experiments versus Taguchi method in the optimization of turning. *Metalurgija* **2011**, *50*, 17–20.

26. Palanikumar, K. Modeling and analysis for surface roughness in machining glass fibre reinforced plastics using response surface methodology. *Mater. Des.* **2007**, *28*, 2611–2618. [CrossRef]

27. Ding, T.; Zhang, S.; Wang, Y. Empirical models and optimal cutting parameters for cutting forces and surface roughness in hard milling of AISI H13 steel. *Int. J. Adv. Manuf. Technol.* **2010**, *51*, 45–55. [CrossRef]

28. Wang, Z.H.; Yuan, J.T.; Liu, T.T.; Huang, J.; Qiao, L. Study on surface roughness in high-speed milling of AlMn1Cu using factorial design and partial least square regression. *Int. J. Adv. Manuf. Technol.* **2015**, *76*, 1783–1792. [CrossRef]

29. Palanisamy, P.; Rajendran, I.; Shanmugasundaram, S. Optimization of machining parameters using genetic algorithm and experimental validation for end-milling operations. *Int. J. Adv. Manuf. Technol.* **2007**, *32*, 644–655. [CrossRef]

30. Zain, A.M.; Haron, H.; Sharif, S. Application of GA to optimize cutting conditions for minimizing surface roughness in end milling machining process. *Expert Syst. Appl.* **2010**, *37*, 4650–4659. [CrossRef]

31. Marler, R.T.; Arora, J.S. The weighted sum method for multi-objective optimization: New insights. *Struct. Multidiscip. Optim.* **2010**, *41*, 853–862. [CrossRef]

32. Xiong, J.; Zhang, G.; Zhang, W. Forming appearance analysis in multi-layer single-pass GMAW-based additive manufacturing. *Int. J. Adv. Manuf. Technol.* **2015**, *80*, 1767–1776. [CrossRef]

*applied sciences*

MDPI

Article

# Thermoelectric Cooling-Aided Bead Geometry Regulation in Wire and Arc-Based Additive Manufacturing of Thin-Walled Structures

Fang Li ⓘ, Shujun Chen *, Junbiao Shi, Yun Zhao and Hongyu Tian

College of Mechanical Engineering and Applied Electronics Technology, Beijing University of Technology, Beijing 100124, China; lif@bjut.edu.cn (F.L.); shibeard@emails.b ut.edu.cn (J.S.); bj_ycw@emails.bjut.edu.cn (Y.Z.); jdthongyu@buu.edu.cn (H.T.)
* Correspondence: sjchen@bjut.edu.cn; Tel.: +86-10-6739-1620

Received: 22 January 2018; Accepted: 29 January 2018; Published: 30 January 2018

**Featured Application: This research helps improve the capability of wire and arc-based additive manufacturing in fabricating thin-walled structures in terms of geometric accuracy, productivity, and microstructure.**

**Abstract:** Wire and arc-based additive manufacturing (WAAM) is a rapidly developing technology which employs a welding arc to melt metal wire for additive manufacturing purposes. During WAAM of thin-walled structures, as the wall height increases, the heat dissipation to the substrate is slowed down gradually and so is the solidification of the molten pool, leading to variation of the bead geometry. Though gradually reducing the heat input via adjusting the process parameters can alleviate this issue, as suggested by previous studies, it relies on experience to a large extent and inevitably sacrifices the deposition rate because the wire feed rate is directly coupled with the heat input. This study introduces for the first time an in-process active cooling system based on thermoelectric cooling technology into WAAM, which aims to eliminate the difference in heat dissipation between upper and lower layers. The case study shows that, with the aid of thermoelectric cooling, the bead width error is reduced by 56.8%, the total fabrication time is reduced by 60.9%, and the average grain size is refined by 25%. The proposed technique provides new insight into bead geometry regulation during WAAM with various benefits in terms of geometric accuracy, productivity, and microstructure.

**Keywords:** additive manufacturing; wire arc additive manufacturing; 3D printing; bead geometry; heat dissipation; thin-walled structure

## 1. Introduction

Additive manufacturing (AM)—the layer-by-layer build-up of parts—has become one of the most promising manufacturing technologies in the past thirty years [1–3]. Parts that are difficult or expensive to fabricate using conventional material removal processes will favor AM. Popular AM processes for metallic materials include power bed fusion, directed energy deposition, etc. [4]. Selective laser melting (SLM) and electron beam melting (EBM) [5], belonging to first category, are relatively superior in terms of geometrical complexity and accuracy. However, they are currently only suitable for fabricating high-value and low-production parts, restricted by high cost and low efficiency.

Wire and arc-based additive manufacturing (WAAM), belonging to the second category, has drawn significant interest from both academia and industry in recent years due to its unique cost and efficiency advantages [6–8]. The low cost is attributed to the easy-to-access wire material and mature welding technologies such as Gas Metal Arc Welding (GMAW). The high efficiency is attributed to its large heat input and high wire feed rate. As reported by Ding et al. [9], its build efficiency can reach up to 50–120 g/min with almost no limitation on the build volume. With these advantages, WAAM is highly competitive in fabricating medium- to large-scale metal parts compared with other AM processes [10], especially for thin-walled structures [11]. Although WAAM has long suffered from low geometry accuracy and poor surface quality resulting from its large heat input, the recent emergence of hybrid manufacturing—integrating additive and subtractive processes into a single setup—may provide a substantial solution to this limitation [12].

Although WAAM originated from welding, their heat dissipation conditions exhibit obvious differences. For welding, the heat is conducted from the molten pool directly to the substrate (see Figure 1a), while for WAAM the heat is dissipated partly to the substrate through previously deposited layers and partly to the ambient air via convection and radiation (see Figure 1b) [13]. With the wall height increasing, the conductive thermal resistance to the substrate is significantly increased and, therefore, an increasing amount of heat is dissipated to the ambient air. However, such a heat dissipation mechanism is less effective than direct conduction to the substrate, which slows down the solidification of the molten pool and therefore leads to a wider and lower weld bead than the expected one. Zhao et al. [14] demonstrated through numerical simulations that the temperature gradient of the molten pool decreases as the wall height increases, and the heat loss quantity descends. Wu et al. [15] presented that the bead geometry varies in the first few layers due to the decreasing cooling rate and gradually becomes steady when the heat input and dissipation reach a balance.

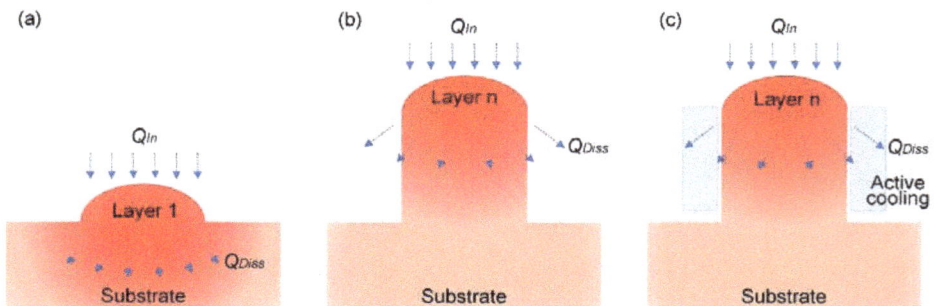

**Figure 1.** Schematic diagram of heat input ($Q_{In}$) and heat dissipation ($Q_{Diss}$) during (**a**) welding; (**b**) wire and arc-based additive manufacturing (WAAM); (**c**) WAAM with active cooling.

Bead geometry is a critical parameter for WAAM that affects geometric accuracy, material utilization, and productivity [16]. A wider bead than the expected one means a larger material removal amount, whereas a lower bead means a longer build time. More seriously, this bead height error will accumulate in the build direction, which implies that the torch-to-workpiece distance will increase continuously, preventing the continuation of the deposition process at upper layers. Therefore, it is crucial to maintain consistent bead geometry during long-term bottom-up deposition. As suggested by previous studies, an effective solution is to gradually reduce the heat input via adjusting the process parameters so as to balance the declining heat dissipation. Wang et al. [17] demonstrated that a 120-layer cylindrical part with good surface morphology could be achieved by reducing the welding current from 140 to 100 A in the first 40 layers with a decreasing step of 1 A per layer and keeping it constant at 100 A for the rest of the layers. Geng et al. [18] pointed out that appropriate interpass temperature control and heat input regulation are effective ways to realize and maintain the consistent thermal boundary condition during a bottom-up additive manufacturing process. A theoretical model was developed to optimize the interlayer temperature and the heat input for each layer deposition. Xiong et al. [19] established a passive vision system for closed-loop control of the layer width and height in the presence of disturbances including the interlayer temperature, heat dissipation condition, and previous forming geometry. It is worth noting that these methods, realized by reducing the heat input, will inevitably sacrifice the deposition rate because the wire feed rate is directly coupled with the heat input. Besides this, real-time adjustment of the heat input relies on experience to a large extent and greatly complicates the build process.

An alternative solution is to regulate the heat dissipation during WAAM, since variations in both heat input and heat dissipation may affect the solidification of the molten pool and, ultimately, the bead geometry. To control the heat dissipation as needed, an in-process active cooling system based on thermoelectric cooling technology is developed in this study. The relatively poor heat dissipation by means of convection and radiation is replaced with strong heat conduction (see Figure 1c), such that the heat dissipation of the upper layers could reach the same level as that of lower layers, thus eliminating their difference in bead geometry with no need for adjusting the process parameters. That is to say, the conflict between consistent bead geometry and high productivity can be resolved. The rest of this paper will introduce the in-process active cooling system and evaluate its efficacy in regulating the bead geometry during WAAM.

## 2. Materials and Methods

### 2.1. Experimental Setup

Figure 2a shows the experimental setup used in this study. It consisted of a robot (RTI2000, IGM, Wiener Neudorf, Austria) equipped with two power supplies (Synergic 5000, Fronius, Mississauga, ON, Canada) to implement WAAM based on Tandem GMAW. In Tandem GMAW, two welding wires that are placed in-line along the travel direction are fed through a single torch [20]. It has been demonstrated to be feasible for AM applications with the benefits of higher productivity than single-wire GMAW. The wire used was 2325 Al alloy (1.2 mm diameter), which was deposited onto a 2219 Al alloy substrate. Pure argon (99.99%) was used as the shielding gas with a constant flow rate of 22 L/min. In the experiments, a thermocouple of type K was located at a depth of 5 mm from the molten pool to record the thermal cycling, a high-speed camera (MotionPro Y4S1, IDT, Tallahassee, FL, United States) was used to capture the molten pool, and a laser scanning confocal microscope (LEXT OLS3100, Olympus, Tokyo, Japan) was used for microstructure analysis.

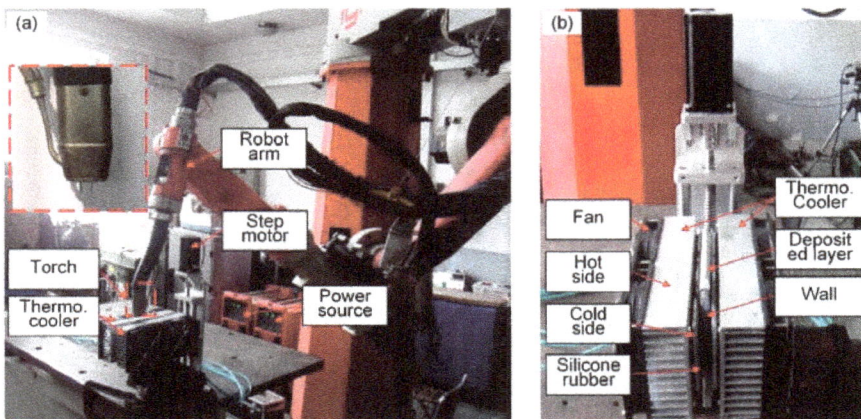

**Figure 2.** Experimental setup: (a) Robot system for WAAM; (b) Enlarged view of the active cooling system.

### 2.2. In-Process Active Cooling System

In order to regulate the heat dissipation of the upper layers during WAAM, it would be more effective if the cooling device was to be directly attached to the side surface instead of to the substrate.

Cooling methods can be classified according to the medium used to transfer the heat. Commonly used methods include air cooling and liquid cooling. However, strong air flow near the molten pool is likely to disrupt the stability of the arc, whereas liquid cooling requires an additional liquid circulation system. Instead, this study adopted thermoelectric cooling technology [21] with the benefits of no circulating liquid, invulnerability to leak, small size, flexible shape, controllable cooling rate, and long life. A thermoelectric cooler operates by the Peltier effect, and is a solid-state active heat pump that transfers heat from one side of the device to the other with consumption of electrical energy. One side gets cooler while the other gets hotter [22]. As shown in Figure 2b, two thermoelectric coolers were distributed symmetrically on the two sides of the wall. The hot side of each cooler was attached to external fans such that it could remain at ambient temperature. The cold side was attached to the side surface of the wall to regulate the heat dissipation, separated by a highly thermally conductive silicone rubber whose function was to ensure good contact between the cooler and the side surface in the presence of the stair-stepping effect. The thickness of the silicone rubber is 2 mm. A step-motor-driven motion system was used to lift the active cooling system in the Z direction along with the increase of the wall height. The rated input voltage of the thermoelectric cooler is 12 V, the nominal cooling power is 180 W, and the maximum temperature difference between the two sides is 63 °C. Generally, the actual cooling power is a function of various factors including ambient temperature, heat sink performance, thermal load, Peltier module geometry, input voltage, etc. The input voltage can be easily adjusted so as to change the cooling power.

## 2.3. Experimental Design

A series of experiments were designed to investigate the respective effects of wall height, heat input, and heat dissipation on the bead geometry, as given in Table 1. These experiments were divided into three groups according to their heat dissipation conditions. Each experiment was repeated three times.

**Table 1.** Heat dissipation conditions and parameter sets in the experiments.

| Run No. | Group No. | Heat Dissipation Conditions | Process Parameters | | | | |
|---|---|---|---|---|---|---|---|
| | | | WFR (m/min) | TS (m/min) | U (V) | I (A) | Cooling Power |
| 1 | | | 3 | 0.3 | 130 | 17.9 | - |
| 2 | I | Flat substrate + conduction | 4 | 0.4 | 176 | 18.6 | - |
| 3 | | | 5 | 0.5 | 222 | 19.5 | - |
| 4 | | | 3 | 0.3 | 130 | 17.9 | - |
| 5 | II | Wall + convection & radiation | 4 | 0.4 | 176 | 18.6 | - |
| 6 | | | 5 | 0.5 | 222 | 19.5 | - |
| 7 | | | 3 | 0.3 | 130 | 17.9 | 240 W |
| 8 | III-1 | | 4 | 0.4 | 176 | 18.6 | 240 W |
| 9 | | Wall + thermoelectric cooling | 5 | 0.5 | 222 | 19.5 | 240 W |
| 10 | | | 3 | 0.3 | 130 | 17.9 | 360 W |
| 11 | III-2 | | 4 | 0.4 | 176 | 18.6 | 360 W |
| 12 | | | 5 | 0.5 | 222 | 19.5 | 360 W |

In Group I, the depositions were performed on a flat substrate with dimensions 150 mm (length) × 150 mm (width) × 6 mm (height) (just like Figure 1a). This could reflect the heat dissipation condition of lower layers. In Group II, the depositions were performed on an existing wall with dimensions 150 mm (length) × 12 mm (width) × 150 mm (height) (just like Figure 1b). The two side surfaces of the wall were exposed to the ambient air to reflect the heat dissipation condition of upper layers. In Group III, the depositions were also performed on an existing wall but the two side surfaces of the wall were attached to the thermoelectric coolers (just like Figure 1c). Groups III-1 and III-2 employed different cooling powers (nominal value) by adjusting the input voltage of the coolers. Within each group, three deposition experiments were conducted with different process parameters. Four main process parameters that affect the heat input, i.e., wire feed rate (*WFR*), travel speed (*TS*), arc voltage (*U*), and arc current (*I*), were considered. The selection of *WFR* covers a wide range. For lower

*WFR*, the welding arc is not quite stable and incomplete melting occurs, while for higher *WFR*, pool overflowing occurs due to excessive heat input as well as the large arc force and strong droplet impingement. Note that the ratio of *WFR* to *TS* is fixed at 10 herein in order to maintain the same area of the bead cross section (i.e., the metal deposition rate per unit length) which is in direct proportion to *WFR/TS* [23]. It is difficult to compare different bead geometries if their cross-sectional areas are not the same. *U* and *I* were obtained from the power supplies once *WFR* and *TS* were determined.

From these experiments, we can analyze (1) the effect of heat input on the bead geometry by comparing the three experiments within each group; (2) the effect of wall height on the bead geometry by comparing Groups I and II; and (3) the effect of heat dissipation on the bead geometry by comparing Groups II and III.

## 3. Results and Discussion

Figure 3a shows some of the deposited layers on the flat substrate, and Figure 3b shows some on the existing wall obtained in these experiments. They all have good bead appearance with little spatters and no visible defects. In the middle segment of each deposited layer, the bead width (*W*) and the bead height (*H*) were measured (see Figure 3c) with the aid of a laser displacement scanner (0.01 mm repeatable precision, HG-C1030, Panasonic, Suzhou, China), and then the ratio of width to height (*RWTH*) was calculated, as given in Table 2. The variation of the bead geometry can be indicated by *RWTH* since the cross-sectional area is a constant value. It can be seen that the three repeated experiments (Samples 1–3) show a maximum error of 0.74 mm in bead width and 0.23 mm in bead height. The discrepancy may be caused by the uncertainty of the molten pool solidification. Only the intermediate values (Sample 2) were recorded for the subsequent analysis.

**Figure 3.** (**a**) Deposited layers on the flat substrate; (**b**) Deposited layers on the existing wall; (**c**) Schematic diagram of bead width and bead height.

**Table 2.** Measurement data.

| Run No. | Sample 1 | | Sample 2 | | Sample 3 | | Maximum Error | |
|---|---|---|---|---|---|---|---|---|
| | Width (mm) | Height (mm) | Width (mm) | Height (mm) | Width (mm) | Height (mm) | Width (mm) | Height (mm) |
| 1 | 9.27 | 3.27 | 9.34 | 3.23 | 9.48 | 3.16 | 0.21 | 0.11 |
| 2 | 9.66 | 3.18 | 9.83 | 3.11 | 10.01 | 3.03 | 0.35 | 0.15 |
| 3 | 10.55 | 3.09 | 10.71 | 3.02 | 10.95 | 2.91 | 0.40 | 0.18 |
| 4 | 10.11 | 2.97 | 10.22 | 2.94 | 10.36 | 2.89 | 0.25 | 0.08 |
| 5 | 11.85 | 2.79 | 12.04 | 2.74 | 12.33 | 2.59 | 0.48 | 0.20 |
| 6 | 12.44 | 2.66 | 12.52 | 2.63 | 12.66 | 2.57 | 0.22 | 0.09 |
| 7 | 9.01 | 3.17 | 9.15 | 3.13 | 9.54 | 3.03 | 0.53 | 0.14 |
| 8 | 10.32 | 2.88 | 10.43 | 2.85 | 10.67 | 2.71 | 0.35 | 0.17 |
| 9 | 11.57 | 2.77 | 11.74 | 2.74 | 12.21 | 2.66 | 0.64 | 0.11 |
| 10 | 8.49 | 3.45 | 8.63 | 3.42 | 8.96 | 3.31 | 0.47 | 0.14 |
| 11 | 9.86 | 3.05 | 9.96 | 3.01 | 10.35 | 2.90 | 0.49 | 0.15 |
| 12 | 10.80 | 2.96 | 11.02 | 2.91 | 11.54 | 2.73 | 0.74 | 0.23 |

Figure 4 compares the bead width, bead height, and *RWTH* between different groups of experiments. Figure 5 compares the bead cross section profiles fitted based on a parabola model [24], which is expressed by

$$Y = -\frac{4H}{W^2}X^2 + H.\qquad(1)$$

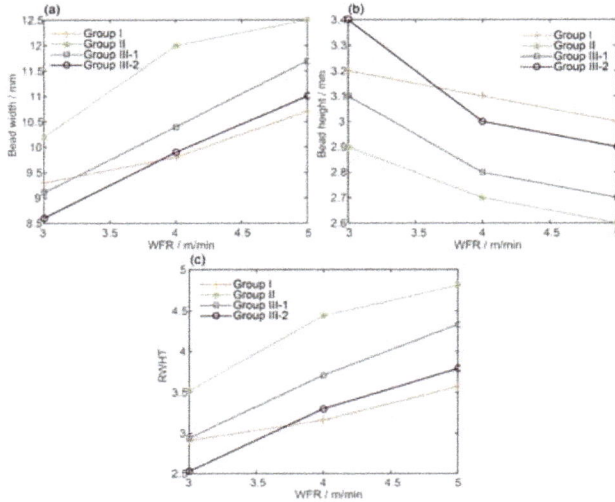

**Figure 4.** (**a**) Relation between bead width and wire feed rate (*WFR*) in Groups I–III; (**b**) Relation between bead height and *WFR* in Groups I–III; (**c**) Relation between ratio of width to height (*RWTH*) and *WFR* in Groups I–III.

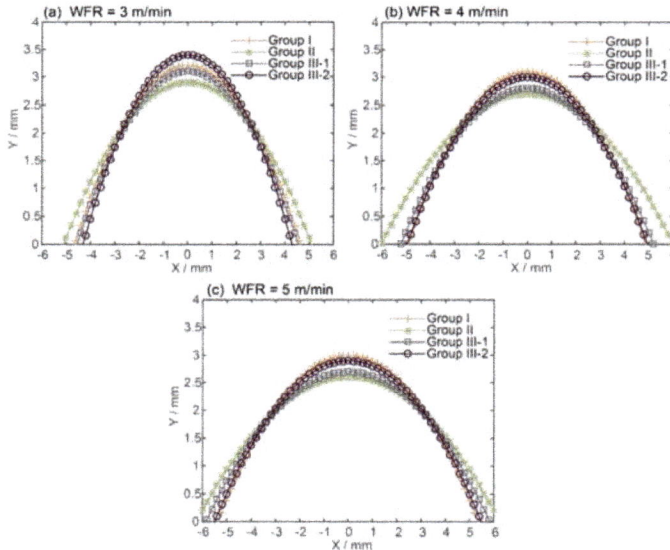

**Figure 5.** Bead cross section profile comparison between Groups I–III (**a**) when *WFR* = 3 m/min; (**b**) when *WFR* = 4 m/min; (**c**) when *WFR* = 5 m/min.

### 3.1. Effect of Heat Input on the Bead Geometry

From Figure 4, it is revealed that within each group of experiments, the bead width tends to be larger and the bead height tends to be smaller (i.e., *RWTH* is larger) as the *WFR* increases. To explain this, the heat input per unit length ($Q_{In}$) is calculated as follows:

$$Q_{In} = \eta \frac{U \times I}{TS} \tag{2}$$

where $\eta$ is the thermal efficiency, which is set to 0.7 in this study [25]. When the WFR is 3 m/min, 4 m/min, or 5 m/min, the corresponding heat input is about $4.65 \times 10^5$ J/m (1630 W), $4.91 \times 10^5$ J/m (2290 W), or $5.19 \times 10^5$ J/m (3030 W), respectively. This clearly shows that the heat input is an increasing function of *WFR*. For smaller *WFR*, i.e., lower heat input, the molten pool does not have enough time to spread before solidification and, therefore, the *RWTH* tends to be smaller. For larger *WFR*, i.e., higher heat input, on the other hand, the viscosity of the material is reduced and the molten pool is much easier to spread [26], leading to larger *RWTH*. The above results prove that adjusting the heat input is one possible way to regulate the bead geometry during WAAM.

### 3.2. Effect of Wall Height on the Bead Geometry

By comparing the *RWTH* between Groups I and II in Figure 4, we can conclude that the upper layers (in Group II) obtain larger *RWTH* than the lower layers (in Group I) under the same process parameters. This is consistent with the above analysis that the heat dissipation condition becomes worse as the wall height increases, thereby slowing down the solidification of the molten pool. Besides this, their difference becomes more noticeable for larger *WFR*. This is because the higher the heat input is, the more difficult the heat dissipation through the upper layers becomes. This explains why it is a great challenge for WAAM to maintain consistent bead geometry from the bottom up. For example, if a 10.5 mm width wall is to be fabricated, the required *WFR* is about 5 m/min for lower layers according to Figure 4a. If this *WFR* remains the same regardless of the increase in the wall height, the obtained bead width is about 12.5 mm at upper layers. Only when the *WFR* is reduced to 3.2 m/min at upper layers can consistent bead geometry be achieved. Nevertheless, it should be noted that the deposition rate (*DR*) is also associated with the *WFR* as follows:

$$DR = \frac{\rho \pi d^2}{4} \times WFR \tag{3}$$

where $\rho$ is the material density and $d$ is the wire diameter. According to this equation, when the *WFR* is reduced from 5 m/min to 3.2 m/min, the corresponding deposition rate is also reduced from 30.5 g/min to 19.5 g/min, which greatly sacrifices productivity. Besides this, this adjustment relies on experience to a large extent and greatly complicates the build process.

### 3.3. Effect of Heat Dissipation on the Bead Geometry

The obtained *RWTH* in Group III-1, as shown in Figure 4c, is obviously decreased compared with that in Group II under the same process parameters, which clearly illustrates that thermoelectric cooling is more effective than convection and radiation in terms of absorbing heat from the molten pool. It is also noteworthy that when *WFR* = 3 m/min, the obtained *RWTH* at upper layers is very close to that at lower layers (in Group I), which implies that the difference in heat dissipation between upper and lower layers is well made up for by thermoelectric cooling. From Figure 5a, we can also see that the two profiles in Group I and Group III-1 tend to overlap. However, for larger *WFR* ranging from 4 to 5 m/min, the effect of thermoelectric cooling is less significant and the obtained *RWTH* is still larger than that at lower layers. This is because the cooling power is limited in this case, and a decreasing proportion of the heat is dissipated through the side surfaces with the increase of the heat input. Besides this, the large arc force and strong droplet impingement resulting from the large heat input also have great effect on the bead geometry, and could not be eliminated by thermoelectric cooling.

When the cooling power is increased to the maximum value in Group III-2, the obtained *RWTH* is further decreased as expected. In this case, the *RWTH* is much closer to that at lower layers when the *WFR* is 4 m/min, but still a little larger when the *WFR* is 5 m/min. If the cooling power is further increased, there is a great possibility to further decrease the *RWTH* to the same level as that at lower layers. To further validate the above results, the thermal cycling and the molten pool shapes of lower and upper layers are also compared (when *WFR* = 4 m/min), as displayed in Figures 6 and 7, respectively. The peak temperature is 313 °C at the lower layer and is 320 °C at the upper layer with thermoelectric cooling, as seen in Figure 6. The small discrepancy (7 °C) may reflect that their heat dissipation conditions are quite similar. If there is no thermoelectric cooling, however, this discrepancy is about 102 °C.

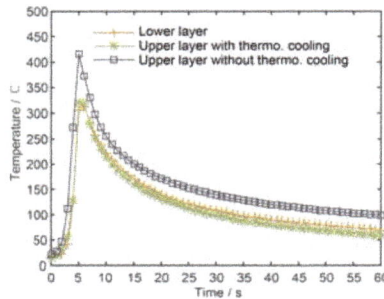

**Figure 6.** Thermal cycling comparison (when *WFR* = 4 m/min).

The molten pool tails at the lower layer and at the upper layer with thermoelectric cooling are also quite similar—like a cone—as seen in Figure 7a,b, respectively. However, the molten pool flows randomly with intensive metallurgical reaction if there is no thermoelectric cooling, which makes it difficult to detect the pool tail as seen in Figure 7c. This further proves that the lower and the upper layers could experience a similar heat dissipation condition with the aid of thermoelectric cooling.

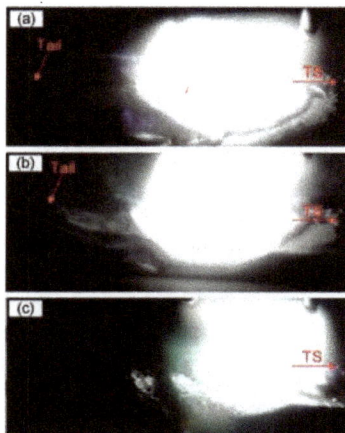

**Figure 7.** Molten pool shape comparison (when *WFR* = 4 m/min): (**a**) at the lower layer; (**b**) at the upper layer with thermoelectric cooling; (**c**) at the upper layer without thermoelectric cooling.

From the above results, we can conclude that once the cooling power matches the heat input, the difference in heat dissipation between the upper and lower layers can be eliminated, which contributes to achieving consistent bead geometry from the bottom up with no need to adjust the process parameters, i.e., without sacrificing the deposition rate. When the heat input is 1630 W or 2290 W, the required cooling power is 240 W or 360 W, respectively, which accounts for 14.8% or 15.7% of the corresponding heat input. If this ratio is smaller, the *RWTH* tends to be larger than that expected; but if larger, the *RWTH* tends to be smaller. This shows that the ratio of cooling power to heat input should be maintained basically constant. It should be pointed out that more accurate adjustment of the cooling power needs further research with regard to thermal modeling of the WAAM process.

## 3.4. Case Study

Two ten-layer deposition cases were compared under the same process parameters: *WFR* = 4 m/min and *TS* = 0.4 m/min. One has the same heat dissipation condition as that in Group III-2 and the other as that in Group II. In the experiments, the interlayer temperature was retained at room temperature to avoid its effect on the bead geometry [15]. Figure 8 shows that the effective bead width and the total wall height are 9.9 mm and 26.8 mm, respectively, with thermoelectric cooling, compared to 12.4 mm and 21.6 mm without thermoelectric cooling. The bead tends to be narrower and taller when the thermoelectric cooling is employed, which is particularly beneficial for the fabrication of thin-walled structures in terms of geometric accuracy and productivity. For example, for when the dimensions of the final part are 100 mm (length) × 8 mm (width) × 100 mm (height), the comparison between the two test cases are given in Table 3. When thermoelectric cooling is employed, the bead width error is reduced by 56.8% and the material utilization is increased by 16.3% because a smaller amount of material needs to be removed through post-processing. On the other hand, the total fabrication time (including both deposition time and cooling time) is reduced significantly by 60.9%. One reason is that the ratio of bead height to bead width is increased and, therefore, the required number of layers is smaller, which means the deposition time is reduced (19.6%). The other reason is that the cooling rate is increased (see Figure 6) and, therefore, the cooling time is reduced (64.4%). It should be noted that without external cooling, the cooling time accounts for the majority of the total time

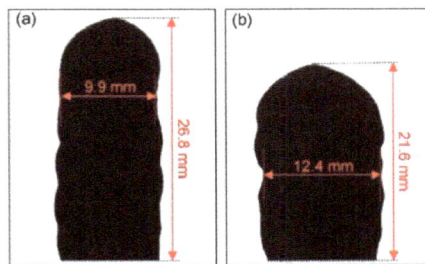

**Figure 8.** Cross section of two ten-layer walls (**a**) with thermoelectric cooling; (**b**) without thermoelectric cooling.

**Table 3.** Comparison between the two test cases.

| Indicator | Convection + Radiation | Thermoelectric Cooling | Relative Error |
|---|---|---|---|
| Bead width error | 4.4 mm | 1.9 mm | −56.8% |
| Material utilization | 64.5% | 80.8% | +16.3% |
| Number of layers | 46 | 37 | −19.6% |
| Deposition time | 690 s | 555 s | −19.6% |
| Cooling time | 8100 s | 2880 s | −64.4% |
| Total time | 8790 s | 3435 s | −60.9% |

The effect of thermoelectric cooling on microstructures was also analyzed. Specimens were taken from both the middle and the bottom parts of the two walls. As shown in Figure 9a,b, the microstructures in the middle parts are characterized by dendrite grains. The average grain size is 5.97 μm with thermoelectric cooling, which is much finer than that without thermoelectric cooling (7.96 μm). This is because the former undergoes a much higher cooling rate that helps to refine the grain size by increasing the number of particles that nucleate grains and by affecting the development of constitutional undercooling [27]. The microstructures in the bottom parts, as shown in Figure 9c,d, are mainly slender columnar grains perpendicular to the substrate as the direction of the grain growth is mainly along the largest temperature gradient [28]. Their difference is not obvious.

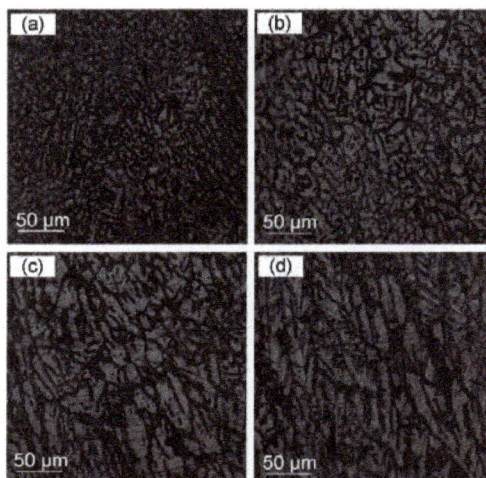

**Figure 9.** Microstructures in the middle parts of the deposited layers (**a**) with thermoelectric cooling; (**b**) without thermoelectric cooling. Microstructures in the bottom parts of the deposited layers (**c**) with thermoelectric cooling; (**d**) without thermoelectric cooling.

## 4. Conclusions

The varying heat dissipation is one of the main constraints of WAAM, degrading geometric accuracy and limiting productivity. Both reducing the heat input via adjusting the process parameters and enhancing the heat dissipation via thermoelectric cooling have the potential to overcome this obstacle. However, the latter exhibits great superiorities as demonstrated in this study. The following conclusions can be drawn from the present study:

(1) The upper and the lower layers could experience a similar heat dissipation condition with the aid of thermoelectric cooling, therefore resulting in similar thermal cycling, molten pool shape, and, ultimately, bead geometry with no need to adjust the process parameters. The case study shows a decrease of 56.8% in bead width error and an increase of 16.3% in material utilization.

(2) The productivity of the WAAM process can be significantly improved with thermoelectric cooling, due to not only the increased ratio of bead height to bead width, but also the reduced interlayer dwell time. The case study shows a decrease of 60.9% in total fabrication time.

(3) Much finer microstructures can be obtained in the middle parts of the deposited walls, and are attributed to the increased cooling rate. The average grain size is reduced from 7.96 μm to 5.97 μm.

In conclusion, the technique offers an innovative way to regulate bead geometry during WAAM, significantly improving its capability in fabricating thin-walled structures in terms of geometric accuracy, productivity, and microstructure. This method can also be extended to other layered manufacturing processes. Future work will focus on the thermal modeling of the WAAM process for accurate adjustment of the cooling power, and improvement of the active cooling system such that it could be fitted to parts with more complex shapes.

**Acknowledgments:** This paper was supported by the National Natural Science Foundation of China (No. 51475009), China Postdoctoral Science Foundation (No. 2017M610726) and Postdoctoral Research Foundation of Chaoyang District (No. 2017ZZ-01-09).

**Author Contributions:** Fang Li developed the experimental setup and wrote the manuscript; Shujun Chen designed the experiments; Junbiao Shi and Yun Zhao performed the experiments; Hongyu Tian analyzed the data.

**Conflicts of Interest:** The authors declare no conflict of interest.

## References

1. Thompson, M.K.; Moroni, G.; Vaneker, T.; Fadel, G.; Campbell, R.I.; Gibson, I.; Bernard, A.; Schulz, J.; Graf, P.; Ahuja, B.; et al. Design for additive manufacturing: Trends, opportunities, considerations, and constraints. *CIRP Ann. Manuf. Technol.* **2016**, *65*, 737–760. [CrossRef]
2. Saboori, A.; Gallo, D.; Biamino, S.; Fino, P.; Lombardi, M. An overview of additive manufacturing of titanium components by directed energy deposition: Microstructure and mechanical properties. *Appl. Sci.* **2017**, *7*, 883. [CrossRef]
3. Gao, W.; Zhang, Y.; Ramanujan, D.; Ramani, K.; Chen, Y.; Williams, C.B.; Wang, C.C.L.; Shin, Y.C.; Zhang, S.; Zavattieri, P.D. The status, challenges, and future of additive manufacturing in engineering. *Comput. Aided Des.* **2015**, *69*, 65–89. [CrossRef]
4. Frazier, W.E. Metal Additive Manufacturing: A Review. *J. Mater. Eng. Perform.* **2014**, *23*, 1917–1928. [CrossRef]
5. Zhong, Y.; Rännar, L.E.; Wikman, S.; Koptyug, A.; Liu, L.; Cui, D.; Shen, Z. Additive manufacturing of ITER first wall panel parts by two approaches: Selective laser melting and electron beam melting. *Fusion Eng. Des.* **2017**, *116*, 24–33. [CrossRef]
6. Ding, D.; Pan, Z.; Cuiuri, D.; Li, H. Wire-feed additive manufacturing of metal components: Technologies, developments and future interests. *Int. J. Adv. Manuf. Technol.* **2015**, *81*, 465–481. [CrossRef]
7. Pan, Z.; Ding, D.; Wu, B.; Cuiuri, D.; Li, H.; Norrish, J. Arc Welding Processes for Additive Manufacturing: A Review. In *Transactions on Intelligent Welding Manufacturing*; Chen, S., Zhang, Y., Feng, Z., Eds.; Springer: Singapore, 2017; Volume 1, pp. 3–24. ISBN 978-981-10-5355-9.
8. Xu, X.; Ding, J.; Ganguly, S.; Diao, C.; Williams, S. Oxide accumulation effects on wire + arc layer-by-layer additive manufacture process. *J. Mater. Process. Technol.* **2017**, *252*, 739–750. [CrossRef]
9. Ding, D.; Pan, Z.; Cuiuri, D.; Li, H. A tool-path generation strategy for wire and arc additive manufacturing. *Int. J. Adv. Manuf. Technol.* **2014**, *73*, 173–183. [CrossRef]
10. Williams, S.W.; Martina, F.; Addison, A.C.; Ding, J.; Pardal, G.; Colegrove, P. Wire + arc additive manufacturing. *Mater. Sci. Technol.* **2016**, *7*, 641–647. [CrossRef]
11. Ding, D.; Pan, Z.; Cuiuri, D.; Li, H. A practical path planning methodology for wire and arc additive manufacturing of thin-walled structures. *Robot. Comput. Integr. Manuf.* **2015**, *34*, 8–19. [CrossRef]
12. Li, F.; Chen, S.; Shi, J.; Tian, H.; Zhao, Y. Evaluation and optimization of a hybrid manufacturing process combining wire arc additive manufacturing with milling for the fabrication of stiffened panels. *Appl. Sci.* **2017**, *7*, 1233. [CrossRef]
13. Michaleris, P. Modeling metal deposition in heat transfer analyses of additive manufacturing processes. *Finite Elem. Anal. Des.* **2014**, *86*, 51–60. [CrossRef]
14. Zhao, H.; Zhang, G.; Yin, Z.; Wu, L. A 3D dynamic analysis of thermal behavior during single-pass multi-layer weld-based rapid prototyping. *J. Mater. Process. Technol.* **2011**, *211*, 488–495. [CrossRef]
15. Wu, B.; Ding, D.; Pan, Z.; Cuiuri, D.; Li, H.; Han, J.; Fei, Z. Effects of heat accumulation on the arc characteristics and metal transfer behavior in wire arc additive manufacturing of Ti6Al4V. *J. Mater. Process. Technol.* **2017**, *250*, 304–312. [CrossRef]
16. Xiong, J.; Zhang, G.; Hu, J.; Wu, L. Bead geometry prediction for robotic GMAW-based rapid manufacturing through a neural network and a second-order regression analysis. *J. Intell. Manuf.* **2014**, *25*, 157–163. [CrossRef]

17. Wang, H.; Jiang, W.; Ouyang, J.; Kovacevic, R. Rapid prototyping of 4043 al-alloy parts by VP-GTAW. *J. Mater. Process. Technol.* **2004**, *148*, 93–102. [CrossRef]
18. Geng, H.; Li, J.; Xiong, J.; Lin, X. Optimisation of interpass temperature and heat input for wire and arc additive manufacturing 5A06 aluminium alloy. *Sci. Technol. Weld. Join.* **2017**, *22*, 472–483. [CrossRef]
19. Xiong, J.; Yin, Z.; Zhang, W. Closed-loop control of variable layer width for thin-walled parts in wire and arc additive manufacturing. *J. Mater. Process. Technol.* **2016**, *233*, 100–106. [CrossRef]
20. Sproesser, G.; Chang, Y.J.; Pittner, A.; Finkbeiner, M.; Rethmeier, M. Environmental energy efficiency of single wire and tandem gas metal arc welding. *Weld. World* **2017**, *61*, 733–743. [CrossRef]
21. Zhao, D.; Tan, G. A review of thermoelectric cooling: Materials, modeling and applications. *Appl. Therm. Eng.* **2014**, *66*, 15–24. [CrossRef]
22. Meng, J.H.; Wang, X.D.; Zhang, X.X. Transient modeling and dynamic characteristics of thermoelectric cooler. *Appl. Energy* **2013**, *108*, 340–348. [CrossRef]
23. Ding, D.; Pan, Z.; Cuiuri, D.; Li, H.; Duin, S.V.; Larkin, N. Bead modelling and implementation of adaptive MAT path in wire and arc additive manufacturing. *Robot. Comput. Integr. Manuf.* **2016**, *39*, 32–42. [CrossRef]
24. Xiong, J.; Zhang, G.; Gao, H.; Wu, L. Modeling of bead section profile and overlapping beads with experimental validation for robotic GMAW-based rapid manufacturing. *Robot. Comput. Integr. Manuf.* **2013**, *29*, 417–423. [CrossRef]
25. Xiong, J.; Zhang, G.; Zhang, W. Forming appearance analysis in multi-layer single-pass GMAW-based additive manufacturing. *Int. J. Adv. Manuf. Technol.* **2015**, *80*, 1767–1776. [CrossRef]
26. Nikam, S.H.; Jain, N.K.; Jhavar, S. Thermal modeling of geometry of single-track deposition in micro-plasma transferred arc deposition process. *J. Mater. Process. Technol.* **2016**, *230*, 121–130. [CrossRef]
27. Easton, M.A.; Stjohn, D.H. Improved prediction of the grain size of aluminum alloys that includes the effect of cooling rate. *Mater. Sci. Eng. A* **2008**, *486*, 8–13. [CrossRef]
28. Cong, B.; Qi, Z.; Qi, B.; Sun, H.; Zhao, G.; Ding, J. A comparative study of additively manufactured thin wall and block structure with Al-6.3%Cu alloy using cold metal transfer process. *Appl. Sci.* **2017**, *7*, 275. [CrossRef]

![applied sciences logo](applied sciences)

MDPI

*Article*

# Online Monitoring Based on Temperature Field Features and Prediction Model for Selective Laser Sintering Process

**Zhehan Chen [1],\*, Xianhui Zong [1], Jing Shi [2] and Xiaohua Zhang [3]**

[1]  School of Mechanical Engineering, University of Science and Technology Beijing, Beijing 100083, China; z1215297392@163.com

[2]  College of Engineering and Applied Science, University of Cincinnati, Cincinnati, OH 45220, USA; shij3@ucmail.uc.edu

[3]  Beijing Materials Handling Research Institute Co., LTD, Beijing 100007, China; zxh_92@163.com

\*  Correspondence: chenzh_ustb@163.com; Tel: +86-138-1174-0959

Received: 13 November 2018; Accepted: 22 November 2018; Published: 25 November 2018

**Featured Application: The researches can bring benefits to those who wants to make the AM process close-loop and make the quality of parts more controllable, not only the additive manufacturing device manufacturers, but also the users of those devices.**

**Abstract:** Selective laser sintering (SLS) is an additive manufacturing technology that can work with a variety of metal materials, and has been widely employed in many applications. The establishment of a data correlation model through the analysis of temperature field images is a recognized research method to realize the monitoring and quality control of the SLS process. In this paper, the key features of the temperature field in the process are extracted from three levels, and the mathematical model and data structure of the key features are constructed. Feature extraction, dimensional reduction, and parameter optimization are realized based on principal component analysis (PCA) and support vector machine (SVM), and the prediction model is built and optimized. Finally, the feasibility of the proposed algorithms and model is verified by experiments.

**Keywords:** temperature field; support vector machine (SVM); process monitoring; quality prediction; selective laser sintering (SLS)

---

## 1. Introduction

Selective laser sintering (SLS) is a powder bed-based additive manufacturing where parts are made directly from three-dimensional CAD data layer-by-layer from the fusion of powder materials [1]. SLS can be used to process a variety of metal materials, and the parts usually have good dimensional accuracy and surface quality [2]. SLS is one of the major additive manufacturing processes that have been widely adopted by industry and investigated by researchers and practitioners. However, physical, chemical, mechanical, and material metallurgical phenomena are extremely complex in the SLS process. It is difficult to develop accurate knowledge about the true state of equipment and the quality of parts during the process. Most often, the stability of the process parameters and the repeatability of the technique cannot be ensured. Therefore, the online prediction of quality issues during the SLS process is extremely challenging, if not impossible. Typically, defects are detected by destructive or non-destructive methods in the as-built parts after the building process is completed, while the causes of the defects are difficult to trace.

In the building process, the temperature field of the working area can quantitatively characterize the real-time states of the metal powder coating and melting pool in time and space, and can indirectly reflect the actual values of the process parameters such as laser power density, scanning track, and scanning speed. According to the characteristics of SLS technology and equipment, an infrared thermal imager is used to collect the temperature field images at any time in the process. By analyzing the temperature field images, the incidence relationship between the temperature field and the states of powder coating and melting pool is established to realize the monitoring and quality control of the SLS process [3,4].

The Doublenskaia team proposed using an infrared thermal imager to acquire image data in the additive manufacturing process, and analyzed the temperature change with time, the state of laser-powder interaction area, and the change of sputtering radiation, and also emphasized the importance of analyzing global temperature data in both time and space dimensions to identify unstable factors in the process [5,6]. The Craeghs team built a temperature measurement and online analysis system for the quality control of selective laser melting. By analyzing the image data of the melting pool, abnormal conditions such as deformation, local overheating, and porosity due to thermal pressure were discovered in advance [7–9]. The Krauss team analyzed the sensitivity of the irradiation level to process parameter fluctuations and porous defects by experimental means, and established a simulation model of the additive manufacturing process based on ANSYS. Macro and micro temperature image features were analyzed by simulation experiments, and the definitions of key features such as the maximum temperature value, thermal diffusivity, and sputtering activity were proposed [10–13].

Many researchers have focused on the features of the melting pool. Bi discussed the effects of the geometric state of parts, laser energy distribution, surface oxidation, and other factors on the temperature of the melting pool [2]. Liu analyzed the effects of the process parameters on laser attenuation and the heating of powder particles by recording the temperature images of the melting pool [14]. Thombansen collected the images of the melting pool by experimental means, and tried to analyze the mutual influence among various factors [15]. Others explored the relationship among process parameters, temperature, and product quality from the aspects of the influence of impurities on product quality and typical abnormal conditions of the process, and subsequently constructed a simulating model of the additive manufacturing process based on the finite element method [16–18].

Existing research demonstrates that the states of the temperature field during the additive manufacturing process are determined by and can reflect the actual values of key process parameters; however, the quantitative relationship between the temperature field and key process parameters is uncertain and difficult to reveal.

In this paper, big data technologies and intelligent algorithms are introduced into the approaches of SLS process monitoring to determine the correlation between the states of the temperature field and key process parameters. This paper aims to predict the process parameters whose variation affects the part quality based on in-process temperature measurement. In Section 2, the key features of the temperature field are proposed; from three aspects, their data structures are defined based on the infrared images of the temperature field. Then, combining principal component analysis (PCA) and a support vector machine (SVM), the prediction model is built, and the modeling process is detailed in Section 3. Finally, the model is trained based on a large amount of experimental data, and is then verified by more experiments.

## 2. Key Features of the Temperature Field

In the SLS process, the temperature field image is collected by an infrared thermal camera, which can quickly acquire temperature images in a certain area at a frequency of about 50 images per second, and yield relevant numerical values, thus providing rich data for process monitoring and quality prediction. Each image can generate a matrix of temperature values corresponding to the state of the SLS process at a certain time. In the continuous production process, thousands of images can be acquired in the SLS process of a single layer of powder coating. When processing the data of the temperature field, reference is made to the image data collected at both the current time and historically.

### 2.1. Definitions of Key Features

Considering the characteristics of the SLS process, the key features of the temperature field are proposed from three aspects: (1) features along the scanning trajectory; (2) features on the single-layer powder coating; and (3) features of the three-dimensional structure. Those key features, as shown in Figure 1, are extracted from the infrared images of the temperature field on the powder layers during the SLS process.

The one-dimensional features mainly include the temperature gradient and cooling rate reflecting the scanning track. The two-dimensional features mainly include the melting pool area, circularity, maximum temperature, heat-affected area information, maximum/average temperature distribution at each point, thermal diffusivity, and sputtering activity. The three-dimensional features mainly refer to the vertical temperature distribution and the three-dimensional reconstruction model based on the maximum or average temperature.

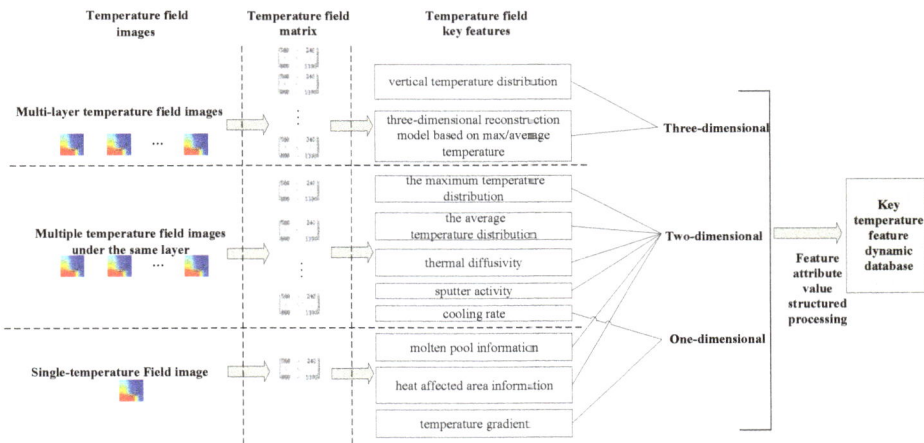

**Figure 1.** Definition of key features of the temperature field.

### 2.1.1. Key Features from a Single Image

The temperature gradient, melting pool information and heat-affected area information is extracted from a single temperature image. The temperature gradient is a physical quantity that describes the rate of temperature change along a specific direction at a certain time on the surface of the powder layer. It mainly reflects the heat distribution of the powder layer after laser scanning.

The melting pool refers to the area where the temperature is greater than the melting point of the material. Melting pool information includes the length, width, area, and circularity. Melting pool geometry information reflects the change of heat input and absorption, which is an important monitoring index of the powder bed fusion processes.

The heat-affected area refers to the area where the temperature on the powder layer is greater than a certain value, and the information of the heat-affected area includes length, width, area and circularity. The heat-affected area has a relatively high temperature, and powder is readily agglomerated, causing the roller to push the part leading to processing failure.

### 2.1.2. Key Features from Multi-Images of the Same Powder Layer

The cooling rate, maximum temperature distribution, average temperature distribution, thermal diffusivity, and sputtering activity are extracted from multiple temperature images of the same layer. The cooling rate refers to the temperature gradient in the concept of time. If it is too high, severe non-equilibrium solidification takes place, while if it is too low, the relevant area is always in a high temperature state, leading to the agglomeration of powders. Both situations affect the quality of the molding parts.

The maximum/average temperature distribution information is formed by the processing results of all of the pixel points. The information is a comprehensive reaction of the current layer's process state, which can describe the temperature statistics of each point in the powder layer. Thermal diffusivity describes the ability of heat diffusion within an object, i.e., the ability of the temperature within the object to tend to be uniform. If the heat accumulation in the powder layer cannot be diffused, it will lead to a high temperature in a small area, and the powder will be agglomerated or burned.

Sputtering refers to the situation in which there is a rapid change of boundary conditions due to rapid energy absorption and diffusion, which leads to the presence of ejecta in the melting pool. Sputtering results in changes in the powder particles' diameter, the thickness of the local powder layer, or the thickness of the cured region. Sputtering particles falling in the uncured region cause an increase in the volume of the melting pool and a sudden change in thermal conductivity, leading to process instability.

### 2.1.3. Key Features from Multi-Images of Multi-Layers

Three-dimensional features based on two-dimensional features are extracted from multi-images of multi-layers, and the three-dimensional structure of the SLS parts is reconstructed for visualization. In addition to using average temperature distribution for temperature-based visual reconstruction, it is also possible to perform visual reconstruction based on other two-dimensional features. The reconstructed three-dimensional structure, where the temperature in a defective place is different from other places and can be distinguished based on color, is an important step in the visual display.

A set of key feature sample data can be extracted at any time during the SLS process. The key feature data of each dimension at different times can provide effective information for reflecting the temperature field change and the state of powder coating and melting pool in the molding process from different angles.

### 2.2. Mathematical Model of Key Features

After a sufficient number of temperature field images are acquired and the key temperature features to be extracted from the temperature field images are determined, the next step is to specify how to extract the relevant temperature features; in other words, to describe the feature extraction method by mathematical expressions.

### 2.2.1. Temperature Gradient

The maximum temperature point in the temperature image is used as the reference point for the temperature gradient, because the maximum temperature point is usually the center point of the melting pool, and the region around it is the area of focus. Taking the reference point as the center, the temperature gradient between the point and the central point is calculated by taking the point along each direction at each interval $s$. Suppose that the central point temperature is $T_0$, and the point temperature is $T_{vs}$; then, the temperature gradient is:

$$\text{grad} = \frac{T_{vs} - T_0}{s}. \tag{1}$$

### 2.2.2. Melting Pool and Heat-Affected Area

According to the temperature image information, a pattern boundary of a melting pool or heat-affected area is obtained. The shape of the melting pool is approximately elliptical, and the shape of the heat-affected area is approximately a long-tailed cone, as illustrated in Figure 2.

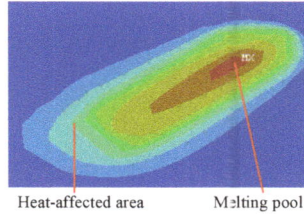

Heat-affected area          Melting pool

**Figure 2.** Melting pool and heat-affected area.

According to the image, circularity is calculated by determining the length, width, area, and perimeter of the ellipse or the long-tailed cone. Circularity refers to the ratio of the area $S_{mea}$ of the current measurement area to the area $S_{circ}$ of a circle with the same perimeter. $S_{circ}$ can be calculated using:

$$S_{circ} = \frac{L^2}{16\pi} \tag{2}$$

where, $L$ is the perimeter of the current ellipse or the long-tailed cone.

The measurement formula for circularity is:

$$Circularity = \frac{S_{mea}}{S_{circ}}. \tag{3}$$

### 2.2.3. Cooling Rate

The cooling rate is calculated by using two adjacent temperature images. The image acquisition interval of the infrared thermal imager is $t_0$. The temperature of the same pixel in different images is $T_{pre}$ and $T_{next}$ The cooling rate of a fixed pixel point on the temperature images of the powder layer can be expressed as:

$$CoolRate = \frac{T_{pre} - T_{next}}{t_0}. \tag{4}$$

### 2.2.4. Max/Average Temperature

The max/average temperature distribution needs to be calculated using all of the temperature images taken by an infrared thermal imager during the fusion of a layer. If the infrared thermal imager takes N temperature images in the printing process of a certain layer, there are N temperature values for a fixed pixel point: $T_0, T_1, \ldots, T_N$. If the maximum temperature distribution is obtained, the characteristic value of the pixel point is:

$$T_{point} = \max(T_0, T_1, \ldots, T_N). \tag{5}$$

If the average temperature distribution is desired, the characteristic value of this pixel point is:

$$T_{point} = \mathrm{mean}(T_0, T_1, \ldots, T_N). \tag{6}$$

### 2.2.5. Thermal Diffusivity

According to the thermal diffusion model, thermal diffusivity can be obtained as follows:

$$a(t) = \frac{1}{t} \left( \frac{1}{T(t) - T_0} \right)^2 \left( \frac{P\varphi}{Vdc_p\rho\sqrt{\pi}} \right)^2 \tag{7}$$

where t is the time variable, $T(t)$ is the temperature at time t, $P$ is the scanning power, $V$ is the scanning rate, d is the scanning interval, $c_p$ is the specific heat capacity, $\rho$ is the material density, and $\varphi$ is the material absorption rate.

### 2.2.6. Sputtering Activity

To quantify the sputtering activity, the sputtering particles in each position of the powder layer are counted as a sputtering activity characteristic value in the position during the printing process.

### 2.3. *Data Structure of Key Features*

Based on the above definition of key features and the mathematical model, the data structure of key features is constructed and displayed based on the sample part with dimensions of 80 mm × 30 mm × 12 mm, as shown in Figure 3.

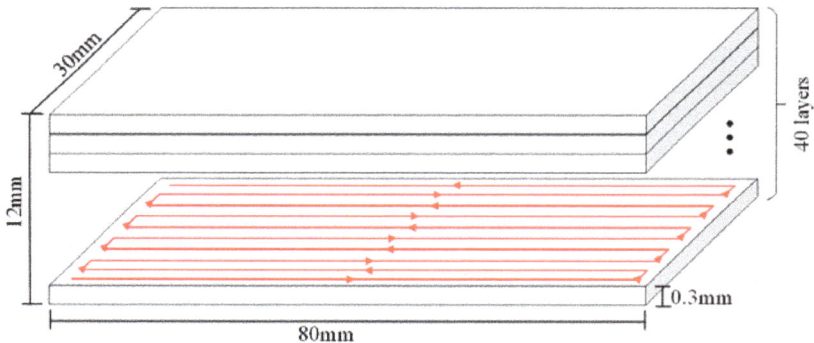

**Figure 3.** Sample parts.

The room temperature is 20 degrees Celsius. The layer thickness is set to 0.3 mm, and 40 layers are printed throughout the process. The infrared thermal imager will collect 1000 images during the sintering process of one single layer, and there will be a total of 40,000 temperature images.

The temperature images are exported as temperature matrixes, and the value of each cell in the matrix represents the temperature of the corresponding location on the image as well as the powder coating. According to the discussion in the previous section, the program based on MATLAB (2016a, MathWorks, Natick, MA, USA) scripts is programmed to process 40,000 temperature distribution images, and the processing results are stored in the key temperature feature database. A total of 40 layers are printed in the experiment. Based on 1000 thermal images taken during the printing process of a certain layer as an example, the temperature value matrix is processed by using the key temperature feature extraction and structured method described in the third section. If extracting information from a single temperature image, a total of 1000 sets of data can be obtained, and the processing result is shown in Figure 4a. The matrix has a total of 1000 columns, and each column is the processing result of a temperature image that can be divided into three parts: temperature gradient, melting pool information, and heat-affected area information. Taking the 1000 temperature images as a group, the cooling rate, max/average temperature distribution, thermal diffusivity, and sputtering activity can be extracted from the group data. Figure 4b is an extraction result of the maximum temperature distribution.

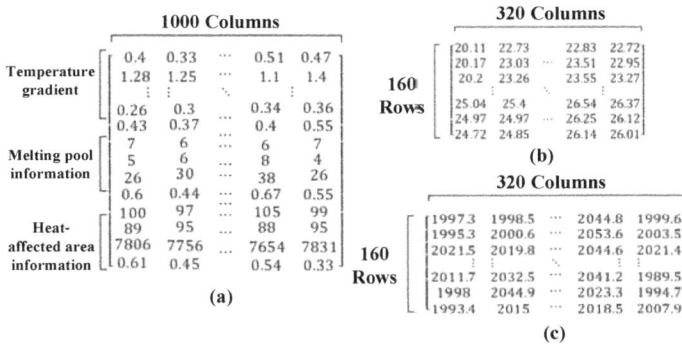

**Figure 4.** Data Structure of Key Features: (**a**) Extracted information from one temperature image; (**b**) The results of the maximum temperature distribution; (**c**) The result of minimum temperature distribution.

The feature extraction of temperature images can not only retain and excavate the potential information, they can also reduce the data quantity, which is beneficial to the storage of data and improves the computation speed of the algorithm.

## 3. Principle of the Prediction Model

During the SLS process, equipment aging, environmental humidity, temperature, and other factors will lead to a difference between the actual value of the process parameters and the set value; fluctuations of the actual value will affect the fusion quality of parts. The essence of quality prediction based on the key feature data of the temperature field is to build a correlation model between the key feature data and the actual values of the SLS process parameters based on many data samples. Furthermore, whether the actual value of a process parameter exceeds the allowable range is determined by monitoring the change of the temperature field, and an alarm is given in time to avoid the waste of materials and time. In this section, the principle of a quality prediction model is described. First, the raw data of key features obtained by an infrared thermal imager are extracted and preprocessed. Then, the PCA method is used to reduce the dimension of features. Finally, the actual value discrimination model of process parameters is established based on a support vector machine (SVM).

### 3.1. Feature Extraction

The key features of the temperature field mainly include the temperature gradient, temperature distribution, melting pool information, and cooling rate. For the specific SLS process, the number of feature values depends on the pixel size of the analyzed temperature field images. Taking $40 \times 80$ images as an example, the key extracted features of the temperature field include:

1.  Temperature gradient information: take the central point of the melting pool as a starting point and calculate the temperature gradient along eight directions—namely, upper, lower, left, right, upper left, upper right, lower left, and lower right—as well as five numerical values in each direction to obtain a total of 40 numerical values.
2.  Temperature distribution information: maximum temperature, minimum temperature, average temperature, and the temperature standard deviation of each row, and column of the images result in $4 \times 80 + 4 \times 40 = 480$ values.
3.  Set three temperature thresholds to record the number of pixel points higher than the threshold.

4. Cooling rate: each pixel point has a cooling rate of $40 \times 80 = 3200$ values.
5. Record the time at which a pixel exceeds the temperature threshold: a total of $3 \times 40 \times 80 = 9600$ values.
6. Record the maximum temperature of the current layer, the length, width, area, and average temperature of the melting pool, the x and y coordinates of the maximum temperature pixel and the current layer number, with a total of eight values.

Therefore, according to images with a size of $40 \times 80$, the key features of the temperature field have a total of 13,331 values.

### 3.2. Dimension Reduction

Since the dimension of the feature vector obtained after feature extraction is large (the feature vector has 13,331 dimensions when the image size is $40 \times 80$), it is not conducive to the establishment of a machine learning model. Therefore, the dimension of the original feature vector is reduced by PCA.

PCA is a statistical method that transforms a set of possible correlation variables into a set of linear uncorrelated variables through orthogonal transformation; the covariance matrix of the original variables is converted into a diagonal matrix. After PCA, a decrease in the feature's quantity can greatly reduce the complexity of the analysis, and more information can be obtained based on fewer features.

There is a certain correlation between the key features of the temperature field in the SLS process. In the original key feature vector, each variable has a specific meaning. After orthogonal transformation, the original feature vector is transformed into a new vector whose components are not related to each other, and that lose the original special meaning. However, from the perspective of information, the new vector retains most of the valuable information contained in the original key feature vector. The importance of each component in the new vector is calculated. Taking the feature vector of 13,331 dimensions extracted based on an image with the size of $40 \times 80$ as an example, the importance of each component after transformation is shown in Table 1.

**Table 1.** Importance of Each Component after Orthogonal Transformation.

| Component Number | Importance |
|:---:|:---:|
| 1 | 58.3% |
| 2 | 73.2% |
| 3 | 80.1% |
| 4 | 84.5% |
| 5 | 86.8% |
| 6 | 89.1% |
| 7 | 90.1% |

As shown in Table 1, the importance of the first three components reaches 80%. To retain more information, we use the feature vector of the first seven features after dimension reduction, the importance of which reaches 90% of the original vector.

### 3.3. Prediction Based on Support Vector Machine (SVM)

In the SLS process, a seven-dimensional vector is obtained after the feature extraction and dimension reduction of each temperature image. There is a correlation between the key features of the temperature field at a certain time, and the actual value of the process parameters at the current time. Therefore, the actual value of the process parameters can be estimated based on the obtained seven-dimensional vector, and the quality of the SLS parts can be predicted. The premise is to establish a correlation model between the seven-dimensional vector and the actual value of the process parameters based on a large quantity of training samples.

Since the process parameters are continuous quantities, to simplify the problem, in this paper, the process parameters are discretized and set to two categories as "out of range" and "appropriate".

*Appl. Sci.* **2018**, *8*, 2383

However, based on the SVM, a classifier is constructed, and many data samples are generated by the SLS simulation model to train the classifier. Finally, the actual value classification prediction model of the process parameters is constructed.

Assume that there are linear separable samples $(x_i, y_i), i = 1, 2, 3, \ldots$, where x is n-dimensional data and y is assumed to be a binary classification whose value is plus/minus one. The optimal classification hyperplane means that the hyperplane can not only separate two kinds of samples correctly—that is, make the training error zero—it can also maximize the classification interval. In d-dimensional space, the linear discriminant function is $g(X) = \omega + b$, and the hyperplane equation is $\omega X + b = 0$. In the formula, $\omega$ is an n-dimensional vector representing the hyperplane's normal, and $b$ is a constant representing a classification threshold. To properly classify all of the samples, the following formula should be satisfied:

$$y_i[(\omega X_i + b)] \geq 1, i = 1, 2, 3, \ldots \tag{8}$$

If the sample is linearly inseparable, there will be some samples that will never meet the above conditions. To balance these classification errors, a relaxation variable is introduced into the constraints. The formula becomes:

$$y_i[(\omega X_i + b)] \geq 1 - \xi_i, i = 1, 2, 3, \ldots \tag{9}$$

When $0 < \xi_i < 1$, sample points are classified correctly. When $\xi_i \geq 1$, the point $X_i$ is incorrectly classified. To solve this problem, a penalty factor is added, and a penalty term $C \sum_{i=1}^{l} \xi_i$ is added to the objective function. The optimization problem can be expressed as:

$$\min \varnothing(\omega, \xi) = \tfrac{1}{2}|\omega| + C \sum_{i=1}^{l} \xi_i$$
$$y_i[(\omega X_i + b)] \geq 1 - \xi_i, i = 1, 2, 3, \ldots \tag{10}$$

$\sum_{i=1}^{l} \xi_i$ represents the upper section of the wrongly classified samples' quantity in the sample set that is used to measure the deviation of sample data from ideal partitioning conditions. C is a penalty factor to control the punishment of the wrongly classified data. When the penalty factor value is small, the classification interval is large, and good generalization performance can be obtained. The above formula can be solved by applying the Lagrangian function to its dual form.

For a nonlinear problem, it can be transformed into the linear data of a high-dimensional space by mapping, and then, linearly separable problems can be solved in a transformed space. According to the correlative functional theories, if the function satisfies Mercer's condition, it corresponds to the inner product of a transformed space, which is called a kernel function $K = (X_i, X_j)$. At present, the commonly used kernel functions include the following: polynomial kernel function, Gaussian kernel function, and sigmoid kernel function.

To prevent the algorithm from being affected by the range of data values resulting in undesired results, the Gaussian kernel function is used in this paper. The Gaussian kernel function, which can map the original data to the infinite dimensional space, has better operation results. It can be expressed as:

$$K(X_i, X) = exp(-gamma \times |X_i - X|^2) \tag{11}$$

## 4. Parameters Optimization

In the process of constructing a classifier model based on SVM, we need to optimize the gamma parameters in the penalty factor C and Gaussian kernel function, and use the meshing search method to determine their values. The basic principle of meshing search is to make C and gamma divide the mesh in a certain range and then traverse all of the points in the mesh to obtain the values. According to a specific combination of C and gamma, the relevant measurement values are calculated and compared, and the optimal combination of parameters is selected as the final parameters.

To train and optimize the parameters of the classifier, a simulation model is established by using ANSYS to generate the training data. As shown in Figure 5, the process prints a total of one layer and 10 lines. Each line will produce 80 groups of temperature data, so a total of 800 temperature field images will be obtained in the simulation process.

In the SLS process, the laser power imposes thermal load on the surface of the powder layer in the form of heat flux, and the spot energy is usually following Gaussian distribution. Due to the reduced thickness of the powder layer, the finite element model uses the two-dimensional (2D) moving Gaussian heat source to simulate the energy input of the laser beam.

$$q = \frac{2AP}{\pi R^2} \exp\left(-\frac{2r^2}{R^2}\right) \tag{12}$$

where R is the laser spot radius, that is, the heat flux density reduced to the heat flux of the center of the spot $\frac{1}{e^2}$ distance from the center of the spot; A is the absorbance of the powder for the laser (wavelength 1064 nm), and the value is 0.77; and r is the distance from the surface of the powder layer to the center of the spot:

$$r = (X - X_0)^2 + (Y - Y_0)^2 \tag{13}$$

Solid–liquid conversion exists in the heating process, and liquid–solid conversion exists in the cooling process. The latent heat of phase change is the heat absorbed or released in the process of phase transformation, which has an important influence on the distribution of the temperature field. The enthalpy can be expressed as:

$$H = \int \rho c dT \tag{14}$$

where $\rho$ is the density of the material, and $c$ is the specific heat of the material.

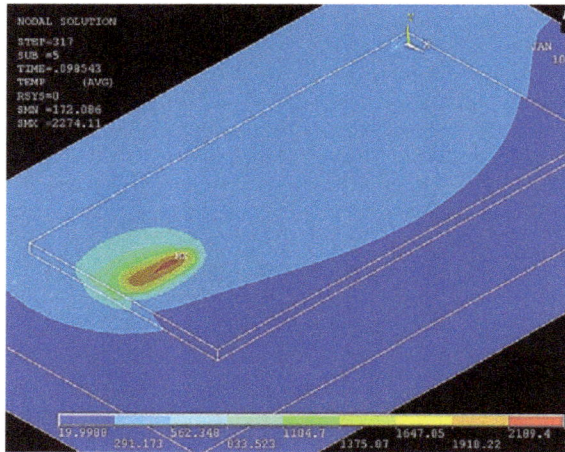

**Figure 5.** Temperature cloud diagram of the finite element simulation process.

The key feature of the temperature field is automatically extracted into a feature $799 \times 13331$ matrix by using a MATLAB script. Each row of the matrix represents the processing result of one temperature image. Since the cooling rate requires the temperature gap of two adjacent images to be divided by the time interval of the obtained images, it is impossible to calculate the cooling rate of the first image so that 799 feature vectors are ultimately obtained. Changing the laser power in the simulation model of the SLS process, the simulation is carried out at 80 W, 100 W, and 120 W, so that 2397 groups of labeled sample data are obtained. It is assumed that in the SLS process, good quality parts with a label of "appropriate" can be obtained when the power is 100 W. The quality of the molding parts cannot meet the production quality requirements when the power deviates from 100 W, and the label of the molding parts is "out of range".

In the experiment, 2397 groups of sample data are divided into a training set and test set. The values of C and gamma are set to range from $10^{-10}$ to $10^{10}$, and each change is 10-fold. For each group of parameters, the training set is used to establish the SVM model for predicting the actual value of the laser power based on the values of the temperature field features, and then, the test set is used to verify and calculate the accuracy and F value.

Determining whether the transient process state is abnormal is a two-category problem, i.e., dividing an instance into positive or negative. For a two-category problem, there are four cases (Table 2). True Positive (TP) represents the correct number in prediction and reality. False Negative (FN) represents a missing report, that is, the number of correct matches that are not found. False Positive (FP) refers to a false alarm, i.e., the given match is incorrect. True Negative (TN) represents the number when both reality and prediction are errors.

**Table 2.** Confusion Matrix.

|  | Prediction True | Prediction False |
|---|---|---|
| True Reality | True Positive (TP) | False Negative (FN) |
| False Reality | False Positive (FP) | True Negative (TN) |

Based on the confusion matrix, it is easy to obtain the evaluation indexes of the two classification algorithms, such as precision, recall, F value, and accuracy. Accuracy is the most common evaluation index, which is the number of correct samples divided by the number of all of the samples. The higher the accuracy, the better the algorithm, which is a very intuitive evaluation index. Nevertheless, sometimes, high accuracy does not mean that an algorithm is good. When the operation set is a deviation class, it will appear that the accuracy is high, but the model is not good. Therefore, it is not feasible to use only accuracy. Accuracy is defined as follows:

$$\text{Accuracy} = (TP + TN)/(TP + FP + TN + FN) \tag{15}$$

Precision (P) and recall (R) are also two metrics that are widely used in statistical theory to evaluate the quality of results. They are defined as follows:

$$P = TP/(TP + FP) \tag{16}$$

$$R = TP/(TP + FN) \tag{17}$$

To evaluate the merits and demerits of different algorithms, the concept of the F value is proposed based on accuracy and recall. The F value considers the P value and the R value synthetically, and when it is high, the algorithm is effective. The F value is defined as follows:

$$F = 2 * P * R/(P + R) \tag{18}$$

In the data used in this experiment, positive samples account for one-third of the data, while negative samples account for two-thirds of the data. The data set has a certain level of deviation, which also makes the F value an important metric.

The top 20 sets of parameter values with the best metrics are shown in Table 3.

**Table 3.** Metrics of Different Parameter Combination Models for the Validation Set.

| Gamma | C | F value | Accuracy |
|-------|---|---------|----------|
| $10^{-1}$ | $10^1$ | 97.9% | 98.6% |
| $10^{-1}$ | $10^3$ | 97.9% | 98.6% |
| $10^{-4}$ | $10^7$ | 97.7% | 98.5% |
| $10^{-1}$ | $10^7$ | 97.7% | 98.5% |
| $10^{-4}$ | $10^9$ | 97.7% | 98.5% |
| $10^{-4}$ | $10^{10}$ | 97.7% | 98.5% |
| $10^{-3}$ | $10^5$ | 97.7% | 98.5% |
| $10^{-2}$ | $10^4$ | 97.7% | 98.5% |
| $10^{-2}$ | $10^5$ | 97.5% | 98.4% |
| $10^{-2}$ | $10^7$ | 97.5% | 98.4% |
| $10^{-2}$ | $10^8$ | 97.5% | 98.4% |
| $10^{-1}$ | $10^5$ | 97.5% | 98.4% |
| $10^{-1}$ | $10^6$ | 97.5% | 98.4% |
| $10^{-3}$ | $10^6$ | 97.3% | 98.3% |
| $10^{-2}$ | $10^6$ | 97.3% | 98.3% |
| $10^{-3}$ | $10^9$ | 97.1% | 98.2% |
| $10^{-3}$ | $10^7$ | 96.9% | 98.0% |
| $10^{-4}$ | $10^6$ | 96.9% | 98.0% |
| $10^{-2}$ | $10^3$ | 96.9% | 98.0% |
| $10^{-1}$ | $10^2$ | 96.9% | 98.0% |

When the gamma is fixed at 0.1, with the change of C, the variation of metrics is shown in Figure 6.

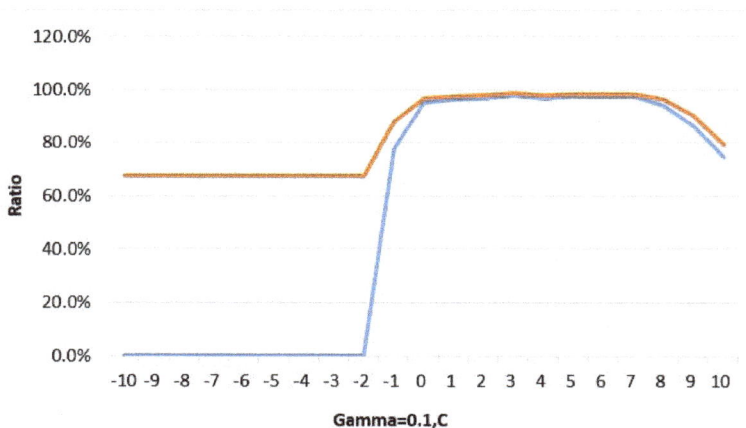

**Figure 6.** Variation curve of metrics with C at gamma = 0.1. (The red curve represents accuracy while the blue curve represents the F value).

As can be seen from the figure, when gamma = 0.1, with the continuous increase in C, both the F value and accuracy have a trend that is first constant, followed by an increase and then a reduction. When C is small, the penalty factor is small. This happens when the penalty term is multiplied by the penalty factor equal to approximately 0, i.e., the penalty factor has little effect on the result of the algorithm model because the value is too small, which also explains why the F value and accuracy are almost constant, even if the penalty factor of the first half is constantly increased. The numerical value of the whole graph shows an initial increase followed by a decrease, indicating that a larger parameter is not better. When the parameter value increases to a certain value, the prediction effect will be reduced. As can be seen from the figure, the initial accuracy is only about 70%, which is merely the proportion of negative samples in the data set. Through verification of the actual value, it is found that when the penalty factor is small, the model will predict all of the samples in the validation set as negative, so the initial accuracy will be maintained at 70%. This also proves that the algorithm has no fault tolerance rate because there is little penalty for the error term in the beginning, i.e., no error is allowed in the training set so that the hyperplane spacing of SVM is very small, resulting in the algorithm prediction results being negative. With the increasing effect of penalty factors, the penalty of error samples increases. By adjusting the proportion of penalty factors to each feature of the model, the prediction accuracy increases. Finally, when the penalty is large enough to blindly pursue the generalization ability, the fault tolerance rate in the training set will be very high and the hyperplane spacing will be very large, so that the validation set will be more prone to error classification, which will lead to a slight decrease in the metrics.

When C = 10, the variation of the metrics with gamma is shown in Figure 7. As can be seen, with the increase in gamma, the metrics also show a tendency to increase first and then decrease. When gamma is very small, the value of the Gaussian kernel function is equal to about 0. Now, SVM has a kernel function whose value is constant zero. The SVM algorithm gives negative prediction results for all of the data, so the initial F value is zero; accuracy is 70% (about equal to the proportion of negative values in the data set). With the increase in gamma, the Gaussian kernel function starts to play a role, which is equivalent to mapping the original data into an infinite dimensional space. However, when the metric increases to a certain value, there will be a reduction. This can be mainly attributed to the over-fitting phenomenon, as the gamma value becomes too large. The phenomenon has a good effect on the training set, but is difficult to generalize to the validation set. Therefore, the metrics increase first and then decrease.

Finally, the parameters of the classifier model are determined as follows: gamma = 0.1, C = 10.

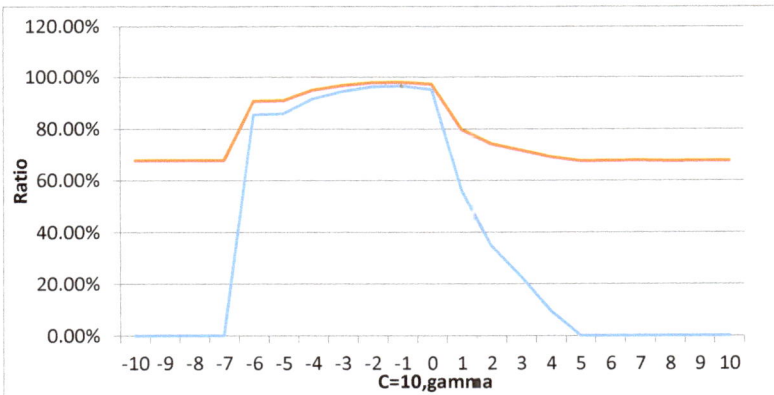

**Figure 7.** Variation curve of metrics with gamma at C = 10. (The red curve represents accuracy, while the blue curve represents the F value).

## 5. Case Study

To verify the feasibility of the proposed prediction model for SLS, experiments are carried out on a SLS machine, and the used material is coated in sand.

The equipment and experimental scene are shown in Figure 8. Three pairs of parts were printed independently with different laser powers. The first pair was printed at a power of 35 W, the second was printed at a power of at 40 W, and the third was printed at a power of at 45 W. During the sintering process of each part, about 40,000 images of the temperature field were produced by the infrared thermal imager. Therefore, there were 80,000 images for each level of laser power.

The used infrared camera is "VarioCAM hr research 480" produced by InfraTec. The camera and its software can measure the surface temperature of the targets in the image area, based on the thermal radiation of those targets. Generally, the thermal radiances are various when referring to different kinds of materials; even for the same material, the thermal radiance will be variable while the temperature of the material changing.

The infrared camera simplified this issue by providing a setup option of the thermal radiances of the target material. In this experiment, we set the thermal radiance of coated sand to be 0.95, which is based on the suggestions from material professionals and some of the literature. Assuming the temperature of the target area is 400 °C, then the energy captured by the camera will be 380 °C; as result of the setup of the thermal radiance, the final temperature shown on the interface of the software will be 400 °C again, which is calculated by 380/0.95.

Actually, the measurement uncertainty exists in any measurement system. The measured temperature values given out by the infrared camera may still not be the real temperatures of the target area. However, the measurement bias will not influence the analysis based on the proposed method, which focuses on the difference between temperatures at different times, instead of the actual values of them.

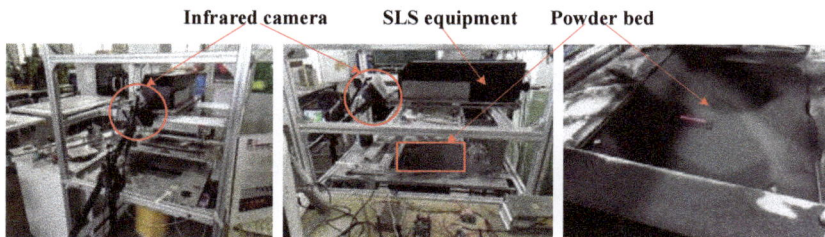

**Figure 8.** Experimental scene of selective laser sintering (SLS).

As illustrated, the key features of the temperature field are extracted from those images, and their values are expressed by matrixes. Then, all 240,000 matrixes are divided into three groups: the training group, validation group, and test group, as shown in Figure 9.

Using the PCA algorithm, the high-dimensional features of the temperature field are reduced to seven features, and the values are represented by an array with seven cells. Some arrays are produced by the SLS process with 35-W and 40-W laser power and are called negative samples; other arrays are produced by the process with 45-W laser power, and they are called positive samples.

Set $Y = 1$ for positive samples and $Y = -1$ for negative samples; then, an SVM model can be built to determine the hyperplane function, which can divide all the arrays into two categories. Finally, for any value of the array, the SVM model can tell us the actual laser power of the process that produces the array; the laser power can be predicted in process based on the key feature values of the temperature field.

Based on the matrixes of the training group, an initial SVM model is built based on the data set of $X_{training}$ and $Y_{training}$; then, the validation group is used to optimize the parameters (gamma and C), using the meshing search method described in Section 4. In this experiment, the optimal value of gamma is 0.1, and the optimal value of C is 10.

Finally, the test group is used to verify the optimized model: the dataset of $X_{testing}$ is input to the established SVM model, and the data set of $Y_{predict}$ is given for each feature array; the evaluating algorithm discussed in Section 4 is run based on $Y_{predict}$ and $Y_{testing}$, with the result showing that the accuracy of laser power prediction is 88.7%, and the F value is 89.3%.

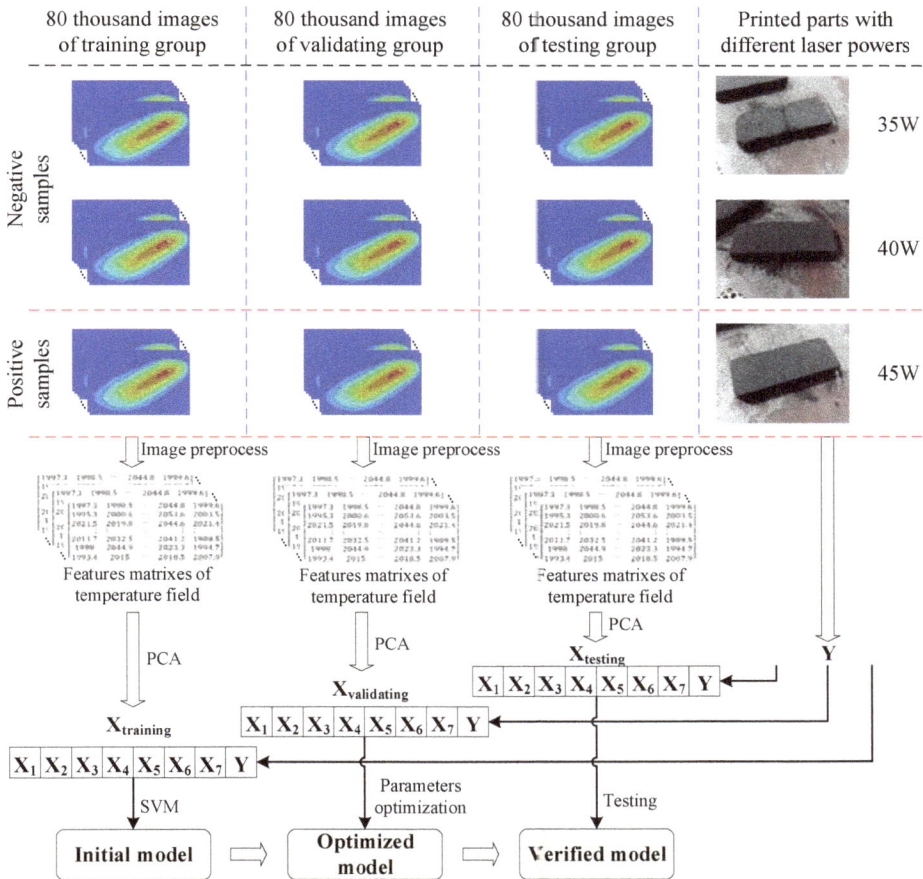

**Figure 9.** Building and verifying process of the predictive model. PCA: principal component analysis.

## 6. Conclusions

In the additive manufacturing processes of powder bed fusion, the temperature of the work space is determined by and reflects the actual values of process parameters, which will result in different quality. To predict the quality of parts based on the temperature field information in SLS, the concept of the temperature field's key features is proposed in this paper. The key feature data are extracted from three aspects: (1) features along with the scanning trajectory; (2) features on the single-layer powder coating; and (3) features of the three-dimensional structure. The mathematical model and data structure of key feature data are established. Based on the PCA method and SVM algorithm, a prediction model based on temperature field images is built after feature extraction, dimension reduction, and parameter optimization. Finally, a practical experiment is carried out to verify the proposed methods. The experimental results show that the optimized prediction model has an accuracy of 88.7% and an F value of 89.3%. It should be noted that the proposed methodology is applicable not only to the SLS process but also to other powder bed fusion processes such as selective laser melting (SLM). In future work, we will continue to explore the correlation between the key features of the temperature field and quality indexes (e.g., porosity), and a real-time analysis model for online monitoring.

**Author Contributions:** Conceptualization, Z.C.; Data curation, X.Z. (Xiaohua Zhang); Formal analysis, Z.C.; Investigation, X.Z. (Xianhui Zong) and X.Z. (Xiaohua Zhang); Methodology, Z.C.; Project administration, Z.C.; Software, X.Z. (Xiaohua Zhang); Supervision, Z.C. and J.S.; Visualization, X.Z. (Xianhui Zong); Writing – original draft, X.Z. (Xianhui Zong); Writing – review & editing, Z.C. and J.S.

**Funding:** This research was funded by National Natural Science Foundation of China, grant number 61803023 and the Fundamental Research Funds for the Central Universities, grant number FRF-TP-16-005A1.

**Conflicts of Interest:** The authors declare no conflict of interest. The funders had no role in the design of the study; in the collection, analyses, or interpretation of data; in the writing of the manuscript, or in the decision to publish the results.

## References

1. Yadroitsev, I.; Krakhmalev, P.; Yadroitsava, I. Selective laser melting of Ti6Al4V alloy for biomedical applications: Temperature monitoring and microstructural evolution. *J. Alloys Compd.* **2014**, *2*, 404–409. [CrossRef]
2. Bi, G.; Sun, C.N.; Gasser, A. Study on influential factors for process monitoring and control in laser aided additive manufacturing. *J. Mater. Process. Technol.* **2013**, *213*, 463–468. [CrossRef]
3. Tapia, G.; Elwany, A. A Review on Process Monitoring and Control in Metal-Based Additive Manufacturing. *J. Manuf. Sci. Eng.* **2014**, *136*, 60801–60811. [CrossRef]
4. Chen, Z.; Zhang, X.; He, K. State space model for online monitoring selective laser melting process using data mining techniques. *Int. J. Manuf. Res.* **2018**, *3*, 270–286. [CrossRef]
5. Bayle, F.; Doubenskaia, M. Selective Laser Melting process monitoring with high speed infra-red camera and pyrometer. *Proc. SPIE Int. Soc. Opt. Eng.* **2008**. [CrossRef]
6. Doubenskaia, M.; Pavlov, M.; Grigoriev, S.; Smurov, I. Definition of brightness temperature and restoration of true temperature in laser cladding using infrared camera. *Surf. Coat. Technol.* **2013**, *220*, 244–247. [CrossRef]
7. Craeghs, T.; Clijsters, S.; Kruth, J.P.; Bechmann, F.; Ebert, M.C. Detection of Process Failures in Layerwise Laser Melting with Optical Process Monitoring. *Phys. Procedia* **2012**, *39*, 753–759. [CrossRef]
8. Craeghs, T.; Clijsters, S.; Yasa, E.; Kruth, J.P. Online quality control of selective laser melting. In Proceedings of the 20th Solid Freeform Fabrication (SFF) Symposium, Austin, TX, USA, 8–10 August 2011.
9. Clijsters, S.; Craeghs, T.; Buls, S.; Kempen, K.; Kruth, J.P. In situ quality control of the selective laser melting process using a high-speed; real-time melt pool monitoring system. *Int. J. Adv. Manuf. Technol.* **2014**, *75*, 1089–1101. [CrossRef]
10. Krauss, H.; Zeugner, T.; Zaeh, M.F. Layerwise Monitoring of the Selective Laser Melting Process by Thermography. *Phys. Procedia* **2014**, *56*, 64–71. [CrossRef]
11. Krauss, H.; Eschey, C.; Zaeh, M.F. Thermography for Monitoring the Selective Laser Melting Process. In Proceedings of the Solid Freeform Fabrication Symposium, Austin, TX, USA, 6–8 August 2012.
12. Schilp, J.; Seidel, C.; Krauss, S.; Weirather, J. Investigations on Temperature Fields during Laser Beam Melting by Means of Process Monitoring and Multiscale Process Modelling. *Adv. Mech. Eng.* **2015**, *6*, 1–7. [CrossRef]
13. Krauss, H.; Zeugner, T.; Zaeh, M. Thermographic process monitoring in powder bed based additive manufacturing. *Rev. Prog. Quantit. Nondestruct. Eval.* **2015**, *1650*, 177–183. [CrossRef]
14. Liu, S.; Farahmand, P.; Kovacevic, R. Optical monitoring of high power direct diode laser cladding. *Opt. Laser Technol.* **2014**, *64*, 363–376. [CrossRef]
15. Thombansen, U.; Abels, P. Observation of melting conditions in selective laser melting of metals (SLM). SPIE LASE. *Int. Soc. Opt. Photonics* **2016**, *9741*. [CrossRef]
16. Kleszczynski, S.; Joschka, J.; Jan, S.; Gerd, W. Error detection in laser beam melting systems by high resolution imaging. In Proceedings of the International Solid Freeform Fabrication Symposium—An Additive Manufacturing Conference, Austin, TX, USA, 6–8 August 2012.
17. Ladewig, A.; Schlick, G.; Fisser, M.; Schulze, V.; Glatzel, U. Influence of the shielding gas flow on the removal of process by-products in the selective laser melting process. *Addit. Manuf.* **2016**, *10*, 1–9. [CrossRef]
18. Zeng, K.; Pal, D.; Stucker, B. A review of thermal analysis methods in laser sintering and selective laser melting. In Proceedings of the Solid Freeform Fabrication Symposium, Austin, TX, USA, 6–8 August 2012.

*applied*
*sciences*

MDPI

Article

# Design & Manufacture of a High-Performance Bicycle Crank by Additive Manufacturing

Iain McEwen [1], David E. Cooper [2], Jay Warnett [3], Nadia Kourra [3], Mark A. Williams [3] and Gregory J. Gibbons [3,*]

[1]  Innovate 2 Make Ltd., Kings Norton Business Centre, Birmingham B30 3HP, UK; iain@mcewenfamily.co.uk
[2]  Progressive Technology Ltd., Hambridge Lane, Newbury RG14 5TS, UK; dave.cooper@progressive-technology.co.uk
[3]  WMG, University of Warwick, Coventry CV4 7AL, UK; J.M.Warnett@warwick.ac.uk (J.W.); N.Kourra@warwick.ac.uk (N.K.); M.A.Williams.1@warwick.ac.uk (M.W.)
*   Correspondence: G.J.Gibbons@warwick.ac.uk

Received: 17 July 2018; Accepted: 7 August 2018; Published: 13 August 2018

**Featured Application: The research presented in this article is focused on the design, manufacture, and validation of an elite bicycle crank.**

**Abstract:** A new practical workflow for the laser Powder Bed Fusion (PBF) process, incorporating topological design, mechanical simulation, manufacture, and validation by computed tomography is presented, uniquely applied to a consumer product (crank for a high-performance racing bicycle), an approach that is tangible and adoptable by industry. The lightweight crank design was realised using topology optimisation software, developing an optimal design iteratively from a simple primitive within a design space and with the addition of load boundary conditions (obtained from prior biomechanical crank force–angle models) and constraints. Parametric design modification was necessary to meet the Design for Additive Manufacturing (DfAM) considerations for PBF to reduce build time, material usage, and post-processing labour. Static testing proved performance close to current market leaders with the PBF manufactured crank found to be stiffer than the benchmark design (static load deflection of 7.0 ± 0.5 mm c.f. 7.67 mm for a Shimano crank at a competitive mass (155 g vs. 175 g). Dynamic mechanical performance proved inadequate, with failure at 2495 ± 125 cycles; the failure mechanism was consistent in both its form and location. This research is valuable and novel as it demonstrates a complete workflow from design, manufacture, post-treatment, and validation of a highly loaded PBF manufactured consumer component, offering practitioners a validated approach to the application of PBF for components with application outside of the accepted sectors (aerospace, biomedical, autosports, space, and power generation).

**Keywords:** additive manufacture; topology optimisation; computed tomography

## 1. Introduction

A high-performance bicycle crank is considered to be one that would be installed on a racing bicycle, the elite of which are operated by professional cycling teams such as those competing in the Tour de France. There has been a considerable effort in developing an understanding of the effect of crank design on the force delivered by the rider [1–5], although there has been very little study of innovation of the crank design itself away from the norm [2]. This is likely in part due to lack of innovation in crank manufacturing methods. The vast majority of common crank designs are effectively box section in their form and are either machined from aluminium in halves then bonded together, or manufactured using Carbon-Fibre-Reinforced Polymer (CFRP) [6] with titanium or aluminium inserts for mechanical interfaces. These methods offer limited opportunity for design innovation.

Component mass is a significant factor in the competitiveness of a racing bicycle and has been driven down to a low level, particularly with the incorporation of CFRP. With cranks, however, stiffness is also a major consideration in order to cycle more efficiently and effectively, maximising performance over both long and short distances through improved power transmission and reduced fatigue. These objectives are suited to topology optimisation (TO), a computer-aided design (CAD) technique. TO has been used previously for designing minimal weight structures [7–9] where low mass is critical to high component performance in the end-use application (e.g., in saving fuel [7,9], increasing stability, and decreasing cost [8]). The effectiveness of TO techniques are often limited by conventional manufacturing techniques due to the complex natural form of the resulting geometry, making the optimal design un-manufacturable.

Additive Manufacturing (AM) introduces an opportunity whereby this theoretically idealised geometry can be manufactured with a greater degree of economic sensibility. TO for AM has therefore become an active research area, utilizing the approach for a number of component performance objectives, such as heat transfer efficiency [10], stiffness [11], and mass [12]. A design produced using this combination of techniques could therefore exhibit maximum stiffness at a minimum mass, achieving ultimate product performance.

The build quality of the complex geometry produced by AM, and the results of functional testing, can both be verified by using Computed Tomography (CT) scanning [13], comparing a CT-generated model with the original CAD data. A similar design investigation has been completed in the high-performance automotive sector, successfully improving the performance of engine valves by designing for AM [14].

Application of this approach to the design and manufacture of an elite bicycle crank to obtain a combination of low mass and high stiffness has not previously been seen. This research paper demonstrates a complete workflow from design, manufacture, post-treatment, and validation of a highly loaded, spiderless, non-drive-side (NDS) crank, offering practitioners a validated approach to the application of Powder Bed Fusion (PBF) to highly loaded components.

## 2. Methodology

The method developed and followed for the design, manufacture, and testing of a high-performance crank is given in Figure 1.

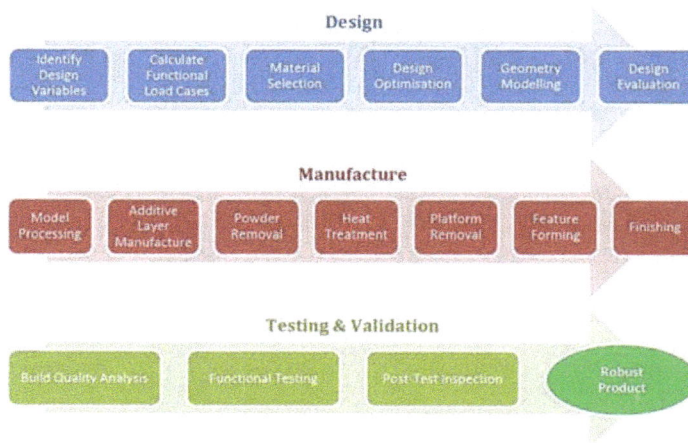

**Figure 1.** Crank Development Methodology.

The optimised design was benchmarked against leading metallic cranks: Hollowtech II [15] and Hollowgram SiSL2 [16], as they offer a similar cost–performance balance.

## 2.1. Design

Crank design began from first principles, optimising for AM whilst maximising stiffness and minimising mass. Ergonomics was considered throughout, with regards to both performance efficiency and comfort.

### 2.1.1. Design Variables

Design variable selection was made based on a common usage and a worst-case scenario. The lower end crank boss was designed to fit a 113 mm square taper cartridge Bottom Bracket (BB), BS 6102-14 (ISO 6696), with 68 mm frame shell width, BS 6102-9 (ISO 6696). A top boss was designed for attachment of the pedal spindle (a 9/16"X 20 TPI British Standard Cycling Thread). The 'crank length', the distance between the pedal centre spindle hole and BB taper, was set as 175 mm. The distance between the outboard face of each crank was set to 147 mm (standard for road products).

### 2.1.2. Boundary Conditions

Both optimisation and Finite Element Analysis (FEA) require accurate input boundary conditions. Crank assemblies must pass a 100,000-cycle fatigue test, alternating loading of each crank (set at 45°) with 1800 N applied 65 mm from its pedal-side face (BS EN ISO 4210-8).

Many studies have developed models to relate crank power to the crank angle. Kautz & Hull [17] defined crank force to be muscular and non-muscular (gravitational or inertial). Bertuccia et al. [18] used horizontal and vertical components of these forces at 70 rpm to produce a force vector profile for realistic loading. The author has scaled this force vector profile to provide the desired 1800 N at 45° to comply with the BS EN ISO 4210 testing conditions. The forces obtained for various crank angles from the resulting force vector profile (Figure 2) agree with those found in numerous studies: maximum load crank angle [19], hip and knee angles of 38°–50° and 73°–145° [20].

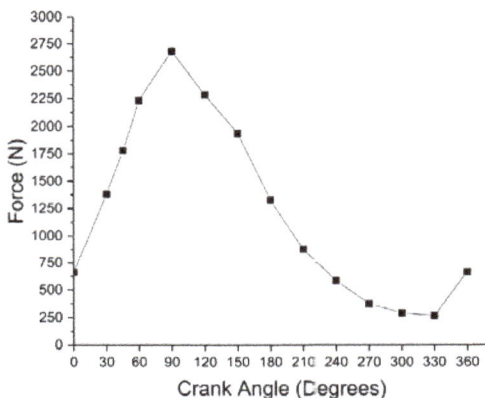

**Figure 2.** Force Applied to Crank at Angle.

The calculated loads from Figure 2 were applied 65 mm from the outboard crank face. An RBE3 (Rigid Body Element) connected this point to the inside face of the pedal spindle hole. This RBE represents the pedal spindle, transferring applied load to a singular external grid to the hole's face. Loads at 30° crank angle intervals were applied to ensure realistic loading and acceptable processing time. Lateral loads were not considered, as with a correctly adjusted Q-Factor, these would be zero.

All degrees of freedom were constrained by a Single-Point Constraint (SPC) at the centre of the square taper void. Each tapered face was connected to this with an infinitely stiff RBE2 element. This infinitely stiff RBE simulates the connection to a BB axle which cannot undergo translational or rotational displacement. While in reality, rotation is allowed about one axis, this produces an unsolvable model for optimisation and this configuration may thus produce an overly stiff result.

### 2.1.3. Material Selection

A high specific stiffness and specific strength is required for high performance at low mass. Therefore density, Young's Modulus, and Yield Strength were identified as key factors. Based on these requirements, Titanium Ti64 and Maraging Steel MS1 are the most favourable options [21] (EOS 1), as shown by the mechanical properties in Table 1.

**Table 1.** EOS Material Properties [22,23].

| Parameter | EOS Ti64 | EOS MS1 |
|---|---|---|
| Density (g/cm$^3$) | 4.41 | 8.05 |
| Young's Mod. (GPa) | $115 \pm 10$ | $180 \pm 20$ |
| Yield Strength (MPa) | $860 \pm 20$ | $1990 \pm 100$ |

### 2.1.4. Optimisation

Topology optimisation was selected as it provides a conceptual geometry for design development. In this case, the design objective was to produce maximum stiffness at minimum mass.

The size of the design space (Figure 3) was limited by physical realities during cycling. Firstly, the limb of the cyclist prevents any protrusions on the pedal-side of the crank so the design space was trimmed in accordance with the Q-Factor and an approximate foot width. The opposite side was made parallel to a hypothetical chain stay to prevent contact with the frame.

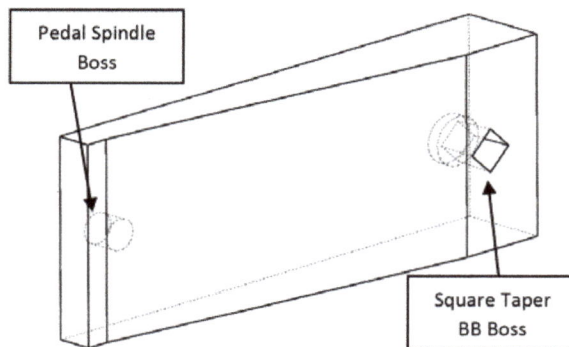

**Figure 3.** Design Space (view from inboard side).

HyperMesh 13.0 (Altair Engineering, Bristol, UK) was used to preprocess and mesh the geometry created in CAD. Genesis Design Studio 14.0 (GRM Consulting, Leamington Spa, UK) was used to perform the geometry optimisation. The Boundary Conditions (Section 2.1.2) were inputs into the model. The mechanical properties of Ti64 were used as the material of choice. Elements of the model were separated into Design Space (DES) (optimised) and Fixed Geometry (FIX) (non-optimised), given as red and blue, respectively. The model configuration can be seen in Figure 4.

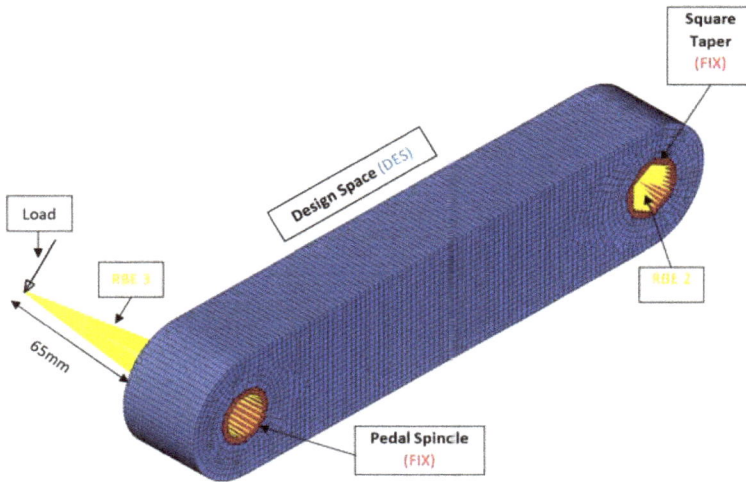

**Figure 4.** Topology Optimisation Model Configuration.

Since AM enables freeform manufacture, no manufacturing constraints were imposed. The design objective was set to minimise strain energy (maximising stiffness). Mass was also defined as a design objective, the 'weighting' of which was increased with progressive iterations.

A relatively coarse initial mesh was used to increase optimisation efficiency. Once reasonable results were obtained, the mesh was refined, increasing the number of tetrahedral elements 10-fold to 504,495 (average element size of 1.09 mm$^2$), increasing the accuracy of the resulting isosurfaces (the surface form of the topology optimisation). Use of tetrahedral elements allowed for more effective packing of the complex solid body. This made an increased mesh resolution possible, which both increases accuracy of results and produces a more precise resulting surface for improved interpretation during CAD modelling.

### 2.1.5. Geometry Creation and Computer-Aided Design

The isosurfaces were exported from Design Studio as an STL file and imported into SolidWorks 2015 (Dassault Systèmes, Vélizy-Villacoublay, France) and directly traced over to produce a practical crank model.

Consideration of AM build direction influenced the development of the design. A vertical build direction starting from the BB boss was selected based on the geometry to reduce the likelihood of manufacturing defects as well as maximise yield.

### 2.1.6. Design Evaluation

FEA was performed in SolidWorks Simulation 2015 (Dassault Systèmes, Vélizy-Villacoublay, France) for each load case; two methods were predominantly used as they would be ultimately replicated by physical testing. The constitutive model for the FEA performed in the SolidWorks static study was a nonlinear elastic model, for both the performance benchmarking and fatigue analysis.

The first test was conducted to provide benchmarking information. Benchmark data for the Hollowtech II and Hollowgram SiSL2 [24] was used, collected by setting the crank at 90°, applying 890 N (225 N preload and 667.5 N main load) 60 mm from the outboard face. This method was repeated 3 times per crank and the average deflection (due to main load) published.

Secondly, the BS EN ISO 4210-8 fatigue test parameters described in Section 2.1.2 were replicated in a static scenario in order to analyse the response of the part under loading.

## 2.2. Manufacture

### 2.2.1. Model Preprocessing

To improve the stability of the build, a solid block base was added to the bottom of the part. The CAD model was exported as an STL file and imported into Magics v19.0 (Materialise Ltd., Sheffield, UK) and support material was generated for the pedal spindle hole as well as the square taper (Figure 5, support material in red). The solid bar was included for calibration of the wire EDM machine for accurate part removal from the build plate.

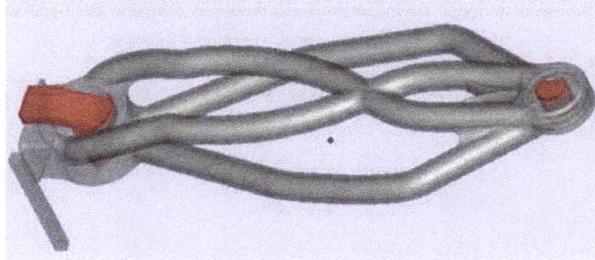

**Figure 5.** Build Model.

### 2.2.2. Additive Manufacturing

Parts were manufactured on an M280 (EOS GmbH, Eschweiler, Germany) in 60 μm layers using standard parameters "Ti64_060_110 Speed" and HSS steel recoater blade. Powder was sieved using the EOS IPCM-M equipment to < 63 μm from a single batch of Ti64 (EOS GmbH, Eschweiler, Germany). The finished build prior to its removal from the machine can be seen in Figure 6.

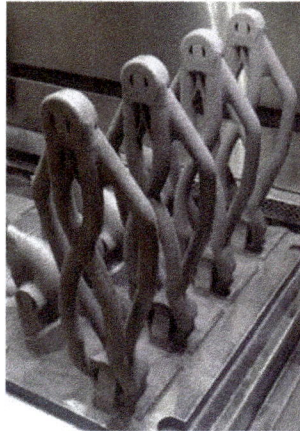

**Figure 6.** Built Parts in Machine.

### 2.2.3. Powder Removal

Once out of the machine, a hydraulic line was attached securely to the build platform to deliver short pulses which produced a vibration in the plate and thus the parts, improving the effectiveness of material removal. The plate was manipulated by hand to ensure that all loose powder had been evacuated.

### 2.2.4. Post-treatment

The parts were heat treated (Tamworth Heat Treatment Ltd., Tamworth, UK) at 800 °C for 4 h in a vacuum [22] to reduce anisotropy [25]. Heat treatment also reduces yield strength, increases elastic modulus, and most importantly, improves fatigue properties.

The components wire EDM were removed from the build platform and the square taper cut into the BB boss (Exetek V500G) (Excetek Technologies Ltd., Shengang, Taiwan). The pedal spindle boss was tapped using an ISO M12 tap. As this was a worst-case scenario design, no surface finishing processes were carried out as this would improve the fatigue characteristics of the components [26].

### 2.3. Validation

### 2.3.1. Post-manufacture Inspection

CT scanning of the part removed from the build plate was completed using an X-Metris TEK XTH 320 LC (Nikon Metrology, Derby, UK). The exposure was set to 300 kV or 40 W for a time of 4 s, a magnification of 1.8× and a gain of 24 dB and a 2 mm Sn filter. The CT scan (3145 projections over 20 h) used a voxel size of 111.11 µm and unsharpness size of 114.05 µm.

The digital 3D model 'actual model' from the scan was compared to the CAD 'nominal model'. The actual and nominal models were aligned using the datum method. The flat faces of the BB boss were used as a datum as they were the most stable during the build and had best flatness, parallelism and dimensional accuracy. A secondary "best fit" method was also used, where an algorithm fitted the models together by identifying the greatest number of common surface points.

### 2.3.2. Functional Validation

Physical testing was conducted to BS EN ISO 4210 by Bureau Veritas UK Ltd., Manchester, UK, using a hydraulic rig to test.

### Static Loading

This test was conducted to benchmark against the Fairwheel Bikes data [24] with the crank set at 90°, 890 N was applied at 65 mm (5 mm further out than the simulation) from the outboard face (Figure 7). A fixed identical crank was placed on the opposite end of the axle to prevent rotation. The magnitude of deflection was manually measured with a steel rule on the crank itself at the pedal spindle boss.

**Figure 7.** Static Test Configuration.

Fatigue

The crank (set at 45°) was loaded with 1800 N 65 mm from its outboard face (Figure 8) at 0.5 Hz (1 s force on; 1 s force off). A solid commercial crank was used to prevent rotation. Two samples from the same PBF build were tested.

**Figure 8.** Fatigue Test Configuration.

## 3. Results and Discussion

### 3.1. Material Selection

Both Ti64 and MS1 exhibit excellent specific properties, however, while the exceedingly high strength of Maraging Steel MS1 may enable a lower mass to be used for a similar performance, its reduced ductility may degrade the fatigue capability of the component. Therefore, Ti64 was selected as the material of choice.

### 3.2. Optimisation

The topology optimisation isosurface results produced are given in Figure 9. Theoretically, this is the ideal design of a solid crank with increasing levels of weight reduction. The complex geometry shown suggests that ALM would be the favoured, if not the only possible method of manufacture for producing this design.

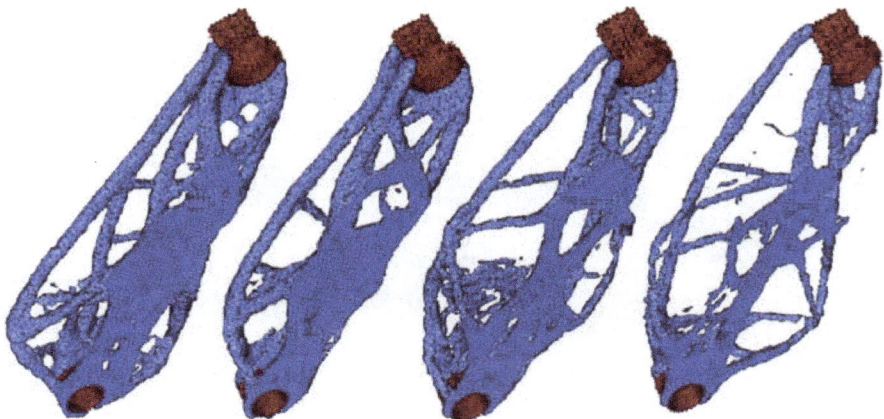

**Figure 9.** Topology Optimisation Isosurface Results (Increasing Mass Reduction Left to Right).

## 3.3. Geometry Creation & Computer-Aided Design

It is clear from Figure 9 that the results from the optimisation could not simply be printed. "Design interpretation" is a key stage of this process; using some of the optimisation results directly may implement too high a level of mass reduction which may not perform functionally. In order to account for this, common features present in all results were identified. These could then be traced in CAD, at which stage each feature could be easily modified using FEA guidance to reach the optimal level of mass reduction. These common features included the 4 primary "limbs" as well as the "auxiliary spars" stemming from these (highlighted yellow and pink, respectively, in Figure 10).

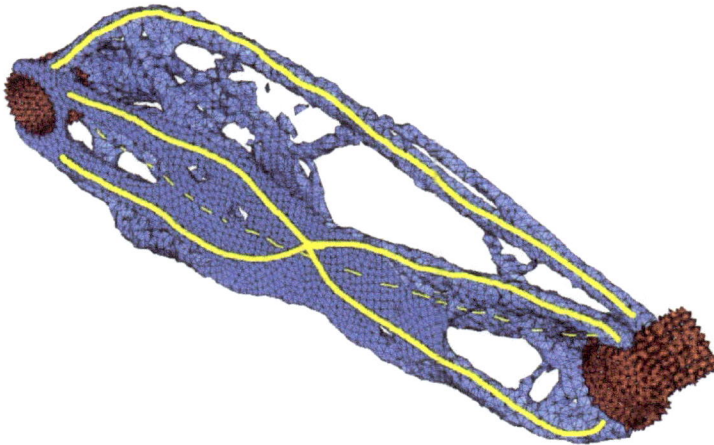

**Figure 10.** Design Interpretation.

The design interpretation produces beam-like structures which can fail through fracture or deflection, determined by their strength and stiffness, respectively. In pure bending, the value for second moment of area (I) is key, where a greater I decreases stress and deflection. Torsion, however, is dependent upon the polar second moment of area (J), where a greater J reduces shear stress and twist. J should be maximised for strength and stiffness [27]. Application of this theory identified three design variables:

1. Tube Diameter 'D': Increased D results in increased J and M
2. Wall Thickness 'WT': Increased WT results in reduced J and increased M
3. Auxiliary Tubing: Addition as per topology results

   D = tube diameter, J = polar second moment of area, M = tube mass, WT = tube wall thickness.

   As this was a feasibility assessment with regards to building the component, a "worst-case" of these variables was used.

### Design for Additive Manufacturing

The current design is produced entirely through focusing on performance; modifications have to be made with regards to Design for Additive Manufacture (DfAM) (Figure 11). Due to its complex geometry, support material would likely be removed by hand and thus be time-consuming and uneconomical. Therefore, all proceeding design work aimed to require no further support material for manufacture (now only required at the pedal spindle, BB taper, and base), reducing waste material, cost, and build time.

A feature was added to support the bottom of the pedal spindle boss (Figure 11). To enable powder removal from inside the part, holes were routed through the component.

The model was heavily filleted both internally and externally to ease stress concentrations and reduce the risk of manufacturing defects. The fillet radius was dependent upon the geometry of the feature, and external radii were matched by corresponding internal fillets to achieve uniform wall thickness throughout.

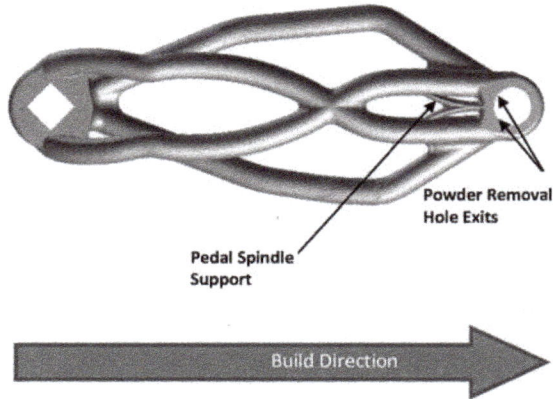

**Figure 11.** Design for Additive Manufacture (DfAM) Diagram.

### 3.4. Design Evaluation

Inconsistencies between the Fairwheel test method [24] and the simulation exist. Fairwheel used a 222.5 N preload before increasing to 890 N, reporting deflection between these two values. The simulated deflection, however, included preload deflection. The simulation used fixed constraints on the square taper faces, whereas in the actual test this was connected to a BB and therefore torsion and bending of this component were not accounted for in the simulated results, whilst approximately 70% of total deflection measured at the pedal came from axle twist [28]. Finally, the Fairwheel test deflection was measured with force applied to the pedal spindle, thus including deformation of the pedal spindle in its measurement. The results are shown in Figure 12a, with a resultant displacement of 2.258 mm.

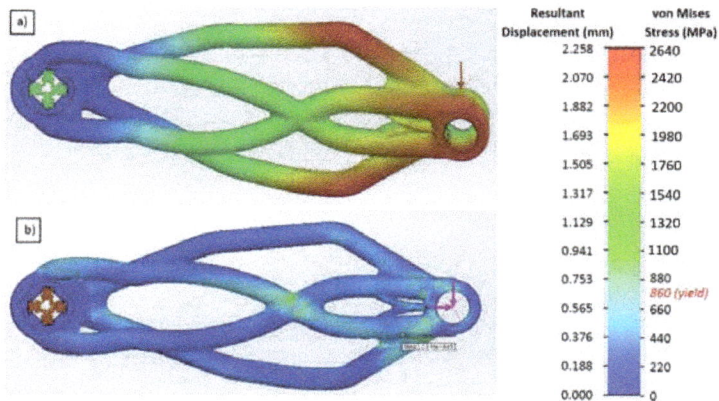

**Figure 12.** (**a**) Static Test Using Fairwheel Bikes Method (Resultant Displacement); (**b**) Static Test Using Fatigue Parameters (von Mises Stress).

The fatigue test was replicated identically in SolidWorks as per BS EN ISO 4210. While this also used fixed constraints on the square taper faces, this would only increase the stress levels and therefore produce a worst-case. Accounting for this, no regions of the results (see Figure 12b) were significantly beyond the yield stress value.

### 3.5. Powder Removal

As this process was performed manually, it was difficult to ensure that all powder was completely removed from the internals of the component. Examination of the CT scanned layers showed no signs of powder remaining within the structure, confirming that powder removal routes were satisfactory.

### 3.6. Post-manufacture Inspection

The datum method showed significant differences at the top of the part. The best fit method, however, produced lower levels of variation (Figure 13). The variance distribution % for both methods is given in Figure 14, where the datum method results in a significant % exceeding 0.5 mm deviation, whereas the best fit does not significantly exceed 0.3 mm.

The maximum geometry variance of under 1 mm is deemed insignificant with regards to its application, and therefore build quality is acceptable. The majority of the deviation is most likely caused by recoater blade contact during the PBF powder layering process as opposed to the effect of residual stresses.

Scans of the internal geometry confirmed that the wall thickness produced is identical to the desired value (Figure 15). Modelling of the pedal spindle support was affected by CT noise reduction algorithms and threshold selection and, due to its smaller wall thickness, geometry comparison of this feature was limited.

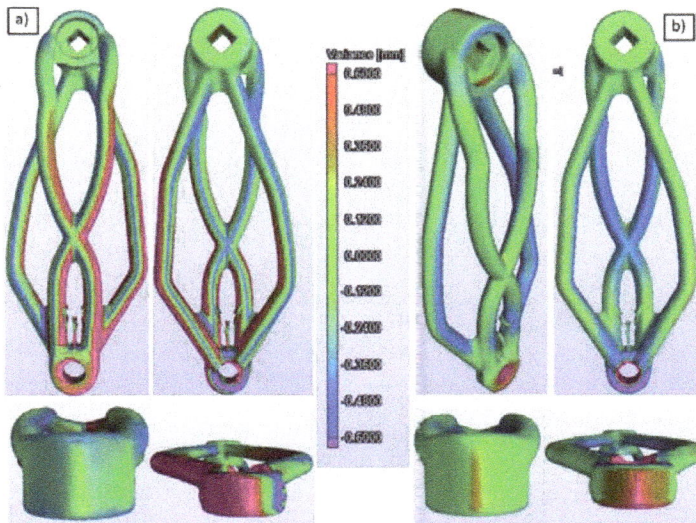

**Figure 13.** Comparison of Actual and Nominal Models; (**a**) Datum-Reference Method; (**b**) Best Fit Alignment Method.

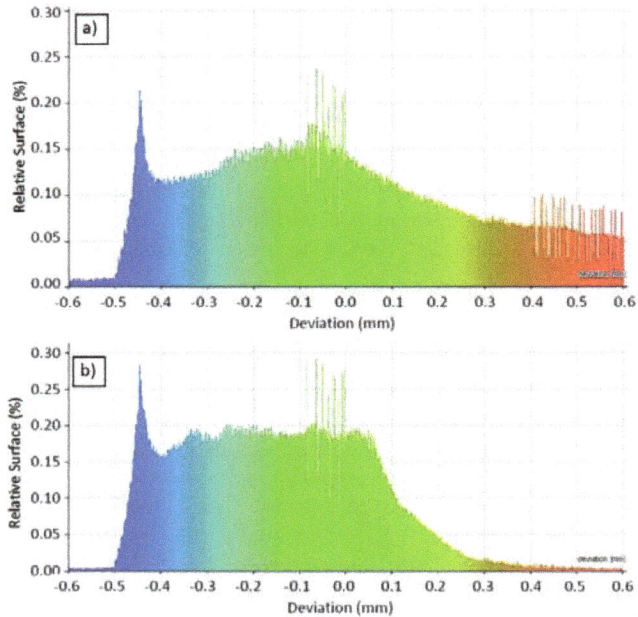

**Figure 14.** Percentage Variation Distribution for (**a**) Datum Method; (**b**) Best Fit Method.

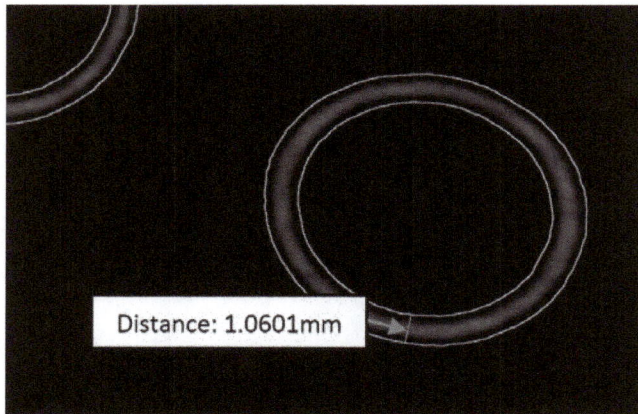

**Figure 15.** CT Scan of Primary Limb Confirming Wall Thickness.

*3.7. Functional Validation*

3.7.1. Static Loading

Under the Fairwheel Bikes method, the crank deflection was $14.0 \pm 0.5$ mm. As an identical crank was used to prevent rotation on the opposite side, it can be assumed that this deformed equally, therefore producing a deflection of $7.0 \pm 0.5$ mm per crank. Twisting of the part could not be accounted for. Lateral displacement occurred at the red patches during simulation seen on the rear main tubes in

Figure 5, suggesting twisting of the crank under load, and therefore this effect was not insignificant but was not being recorded in its own right.

The simulation of this method produced a deflection of just 2.258 mm (c.f. 7.0 ± 0.5 mm). It is likely that much of this deviation was due to the simulation not measuring torsion of the BB. This is reinforced by claims that roughly 70% of deflection was due to BB torsion [29], with the simulated value of 2.258 mm being 32% of the measured value.

A comparison of this design with the benchmark cranks is given in Table 2 and indicates that this worst-case design for the AM crank is stiffer than current market leading designs, and at a competitive weight.

**Table 2.** Benchmarked Results.

| Parameter | Shimano | Cannondale | AM |
|:---:|:---:|:---:|:---:|
| Mass (g) | 175 | 125 | 155 |
| Deflection (mm) | 7.67 * | 6.91 * | 7.0 ± 0.5 |

* 50–200 lb distance only [24].

3.7.2. Fatigue

Both of the tests completed failed catastrophically, through identical mechanisms, after similar durations (2370 and 2620 cycles) (Figure 16). During the design phase, yield stress (860 MPa) had been used as the mechanical limit, whereas the actual capability of a part under fatigue is lower than this. Rekedal [30] investigated high cycle fatigue life of Ti64 using identical build and heat treatment parameters to those here. The Rekedal study suggests that peak stress below 250 MPa would be required to achieve 100,000 cycles. Edwards and Ramulu [25] found that we would expect failure earlier than when it physically occurred.

The fatigue configuration was simulated in a static test (Figure 12b). At the centre section where critical failure occurred, a higher stress can be seen (green). Combining this with the components post-fatigue (Figure 16), it is likely that the fracture initiated at this location and propagated across the part under cyclic loading. The other noncritical area of damage (Figure 16) is also a region of higher stress. The area located near the pedal spindle, however where the maximum stress is identified, showed no signs of external damage, although this is an area of high material concentration and therefore had better stress-relieving characteristics.

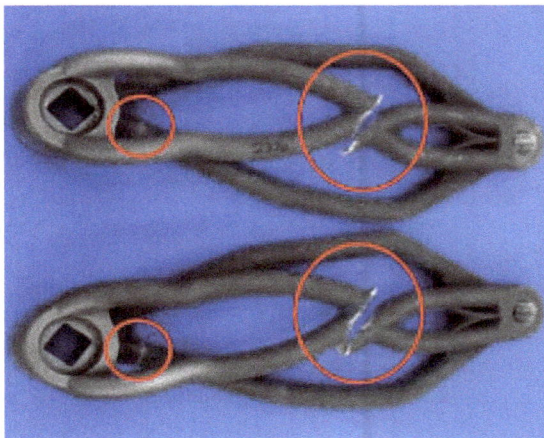

**Figure 16.** Damage Identification from Fatigue; Left: N = 2370, Right: N = 2620.

## 4. Conclusions

It is expected that the UK's High Value Manufacturing sector can capture over £3.5 bn GVA per year for the UK of the rapidly growing global market for AM products and services by 2025 [31]. Despite the exciting potential and progress to date, many UK companies, especially within the SME community, lack the awareness, resources, or confidence to apply AM as a core and integral part of their manufacturing toolkit. Lack of design capability and the knowledge required to take full advantage of the benefits offered by AM were highlighted in this recent UK Government report [31].

The research presented in this paper provides a methodology through which a product can be fundamentally redesigned, manufactured using state of the art AM metals technology, and validated, an approach that can be adopted by industry to engage with this technology and harness the benefits. A novel aspect of this research lies in the product focus; unusually, this research is not aimed at very high value products, such as for aerospace [32] or biomedical [33] applications, but is focused on consumer market product, making the approach more tangible for industry, and in particular, for the SME sector, which accounts for 60% of the UK's manufacturing output.

A further novelty of this research is that it describes the product realisation approach across design, manufacture, and validation, rather than exploring these as individual aspects. In developing and demonstrating this process flow, we have demonstrated the efficacy of PBF, particularly when designing specifically towards using this as a manufacturing tool. Topology optimisation was free from conventional manufacturing constraints to profit from a fully maximised design space, enabling the design of a crank that was stiffer than the benchmark design (static load deflection of $7.0 \pm 0.5$ mm c.f. 7.67 mm for a Shimano crank at a competitive mass (155 g vs. 175 g).

The successful use of CT scanning as a method of post-manufacture inspection proved an acceptable build quality of a tall, complex design, and that powder removal was complete such that the physical part represented the digital model. Nonlinear elastic simulation could then be used to accurately model component performance, as well as diagnose catastrophic failure of the part under fatigue. The physical testing validated both simulation studies and gives confidence that it provides an accurate representation of a component produced by AM.

The value of this research is in the demonstration of the realisation of a design through topology optimisation which is unlikely to have been produced by human invention. The computational nature of topology optimisation reduced manual analysis and an iterative design optimisation process. Its incorporation increased both the efficiency and effectiveness of the design phase significantly, whilst increasing functional performance to a degree that may have been unachievable without its inclusion. Manual design for AM was required, however, with the addition of features purely to enable a successful build. This might suggest that a limitation of topology optimisation is that it is almost too free, and that the development of design for AM considerations would be beneficial.

Furthermore, incorporation of biomechanics specific to the realistic loading conditions enabled the production of a product which has the potential to outperform more conventional designs.

The results of this research also saw physical fatigue testing which surpassed the lifetime expected for titanium components produced by PBF [25]. This might suggest that technology advancements are seeing the quality of PBF components improve and that further research into the quality and integrity of the resulting specimens may need to be maintained, where improved mechanical properties would enable the design capability to be pushed further.

Further value lies in the quality of the resulting component itself, whereby this worst-case scenario design performed within the region of existing market leader products, design development of which will likely see their performance surpassed. This further reinforces the success of this optimised process for design, manufacture, and validation of components manufactured by PBF using the techniques demonstrated here. Again, this provides industry with confidence in the PBF process for application to highly loaded components, ultimately moving towards the AM-UK National Strategy [31] goal of increasing awareness and confidence in AM as a manufacturing tool.

**Author Contributions:** Conceptualization, G.J.G. and David E.C.; Methodology, I.M., D.E., N.K., and J.W.; Software, I.M., N.K., and J.W.; Validation, I.M., D.E.C., N.K., and J.W.; Formal Analysis, I.M., N.K., and J.W.; Investigation, I.M., D.E.C., N.K., and J.W.; Resources, G.J.G., and M.A.W.; Data Curation, G.J.G.; Writing-Original Draft Preparation, I.M., D.E.C., J.W.; Writing-Review & Editing, G.J.G.; Visualization, I.M., N.K., G.J.G.; Supervision, G.J.G., and M.A.W.; Project Administration, G.J.G. and M.A.W.; Funding Acquisition, G.J.G.

**Funding:** This research received no external funding.

**Acknowledgments:** The authors wish to thank the High Value Manufacturing Catapult for supporting this research.

**Conflicts of Interest:** The authors declare no conflict of interest.

## References

1. Casas, O.V.; Dalazen, R.; Balbinot, A. 3D Load Cell for Measure Force in a Bicycle Crank. *MEAS* **2016**, *93*, 189–201. [CrossRef]
2. Zamparo, P.; Minetti, A.E.; di Pramperoa, P.E. Mechanical Efficiency of Cycling with a New Developed Pedal–crank. *J. Biomech.* **2002**, *35*, 1387–1398. [CrossRef]
3. Yoshihuku, Y.; Herzog, W. Optimal Design Parameters of the Bicycle-rider System for Maximal Muscle Power Output. *J. Biomech.* **1990**, *23*, 1069–1079. [CrossRef]
4. Hull, M.L.; Gonzalez, H. Bivariate Optimization of Pedalling Rate and Crank Arm Length in Cycling. *J. Biomech.* **1988**, *21*, 839–849. [CrossRef]
5. Gross, V.J.; Bennett, C.A. Bicycle Crank Length. In Proceedings of the 6th International Congress of the International Ergonomics Association "Old World, New World, One World" and Technical, Programme of the 20th Annual Meeting of the Human Factors Society, College Park, MD, USA, 11–16 July 1976; pp. 415–421. [CrossRef]
6. Chang, R.R.; Dai, W.J.; Wu, F.Y.; Jia, S.Y.; Tan, H.M. Design and Manufacturing of a Laminated Composite Bicycle Crank. *Procedia Eng.* **2013**, *67*, 497–505. [CrossRef]
7. Dima, G.; Balcu, I.; Zamfir, M. Method for Lightweight Optimization for Aerospace Milled Parts—Case Study for a Helicopter Pilot Lightweight Crashworthy Seat Side Struts. *Procedia Technol.* **2015**, *19*, 161–168. [CrossRef]
8. De Souza, R.; Miguel, L.F.F.; Lopez, R.H.; Torii, A.J. A Procedure for the Size, Shape and Topology Optimization of Transmission Line Tower Structures. *Eng. Struct.* **2016**, *111*, 162–184. [CrossRef]
9. Das, R.; Jones, R. Topology Optimisation of a Bulkhead Component used in Aircrafts Using an Evolutionary Algorithm. *Proc. Eng.* **2011**, *10*, 2867–2872. [CrossRef]
10. Dede, E.M.; Joshi, S.N.; Zhou, F. Topology Optimization, Additive Layer Manufacturing, and Experimental Testing of an Air-Cooled Heat Sink. *J. Mech. Des.* **2015**, *137*, 1–9. [CrossRef]
11. Takezawaa, A.; Yonekura, K.; Koizumi, Y.; Zhang, X.; Kitamura, M. Isotropic Ti–6Al–4V Lattice via Topology Optimization and Electron-beam Melting. *Addit. Manuf.* **2018**, *22*, 634–642. [CrossRef]
12. Barbieri, S.G.; Giacopini, M.; Mantovani, V.M.S. A Design Strategy Based on Topology Optimization Techniques for an Additive Manufactured High Performance Engine Piston. *Procedia Manuf.* **2017**, *11*, 641–649. [CrossRef]
13. Thompson, A.; Senin, N.; Maskery, I.; Körner, L.; Law, S.; Leach, R. Internal Surface Measurement of Metal Powder Bed Fusion Parts. *Addit. Manuf.* **2018**, *20*, 126–133. [CrossRef]
14. Cooper, D.; Thornby, J.; Blundell, N.; Henrys, R.; Williams, M.A.; Gibbons, G. Design and Manufacture of High Performance Hollow Engine Valves by Additive Layer Manufacturing. *JMAD* **2015**, *69*, 44–55. [CrossRef]
15. Hollowtech II Crankset. Available online: https://bike.shimano.com/en-EU/technologies/component/details/hollowtech-2.html (accessed on 16 July 2018).
16. Hollowgram SiSL2 Road. Available online: www.cannondale.com/en/USA/Gear/GearDetail?Id=cbde5cb1-afbe-4b34-b6ff-3c11b51c8870 (accessed on 16 July 2018).
17. Kautz, S.A.; Hull., M.L. A Theoretical Basis for Interpreting the Force Applied to Pedal in Cycling. *J. Biomech.* **1993**, *26*, 55–165. [CrossRef]

18. Bertuccia, W.; Grappea, F.; Girarda, A.; Betikab, A.; Rouillonc, J.D. Effects on the Crank Torque Profile when Changing Pedalling Cadence in Level Ground and Uphill Road Cycling. *J. Biomech.* **2005**, *38*, 1003–1010. [CrossRef] [PubMed]

19. Bini, R.R.; Humea, P.A.; Cerviric, A. A Comparison of Cycling SRM Crank and Strain Gauge Instrumented Pedal Measures of Peak Torque, Crank Angle at Peak Torque and Power Output. *Procedia Eng.* **2011**, *13*, 56–61. [CrossRef]

20. Human Performance Capabilities. Available online: https://msis.jsc.nasa.gov/sections/section04.htm#_4.9_STRENGTH (accessed on 16 July 2018).

21. Materials for Metal Manufacturing. Available online: www.eos.info/material-m (accessed on 16 July 2018).

22. Ti64 Material Data Sheet. Available online: https://cdn.eos.info/a4eeb73865d54434/5926811b3739/Ti-Ti64_9011-0014_9011-0039_M290_Material_data_sheet_11-17_en.pdf (accessed on 26 July 2018).

23. MS1 Material Data Sheet. Available online: www.eos.info/material-m/download/material-datasheet-eos-maragingsteel-ms1.pdf (accessed on 26 July 2018).

24. Road Bike Crank Test. Available online: https://blog.fairwheelbikes.com/reviews-and-testing/road-bike-crank-testing (accessed on 16 July 2018).

25. Edwards, P.; Ramulu, M. Fatigue Performance Evaluation of Selective Laser Melted Ti–6Al–4V. *Mater. Sci. Eng. A* **2014**, *598*, 327–337. [CrossRef]

26. Greitemeier, D.; Dalle Donne, C.; Syassen, F.; Eufinger, J.; Melz, T. Effect of Surface Roughness on Fatigue Performance of Additive Manufactured Ti–6Al–4V. *Mater. Sci. Technol.* **2013**, *53*, 629–634. [CrossRef]

27. Budynas, R.G.; Nisbett, K.J. *Shigley's Mechanical Engineering Design*; McGraw-Hill: London, UK, 2011.

28. BikeRadar: Complete Guide to Bottom Brackets. Available online: www.bikeradar.com/gear/article/complete-guide-tobottom-brackets-36660 (accessed on 13 April 2014).

29. Huang, J. BikeRadar: Complete Guide to Bottom Brackets. Available online: www.bikeradar.com/gear/article/complete-guide-to-bottom-brackets-36660/ (accessed on 1 March 2018).

30. Rekedal, K.D. Investigation of the High-Cycle Fatigue Life of SLM and HIP Ti–6Al–4V. Master's Thesis, Ohio Air Force Institute of Technology, Wright-Patterson AFB, OH, USA, 2015.

31. Additive Manufacturing UK National Strategy 2018-25. Available online: https://am-uk.org/project/additive-manufacturing-uk-national-strategy-2018-25/ (accessed on 23 May 2018).

32. Bici, M.; Brischetto, S.; Campana, F.; Ferro, C.G.; Seclì, C.; Varetti, S.; Maggiore, P.; Mazza, A. Development of a Multifunctional Panel for Aerospace use Through SLM Additive Manufacturing. *Procedia CIRP* **2018**, *67*, 215–220. [CrossRef]

33. Martorelli, M.; Maietta, S.; Gloria, A.; De Santis, R.; Pei, E.; Lanzotti, A. Design and Analysis of 3D Customized Models of a Human Mandible. *Procedia CIRP* **2016**, *49*, 199–202. [CrossRef]

applied
sciences

MDPI

*Article*

# Comparison of Laser-Engraved Hole Properties between Cold-Rolled and Laser Additive Manufactured Stainless Steel Sheets

**Matti Manninen [1,\*], Marika Hirvimäki [1], Ville-Pekka Matilainen [1] and Antti Salminen [1,2]**

[1]   Laboratory of Laser Processing, School of Energy Systems, Lappeenranta University of Technology, 53850 Lappeenranta, Finland; marika.hirvimaki@lut.fi (M.H.); ville-pekka.matilainen@lut.fi (V.-P.M.); antti.salminen@lut.fi (A.S.)
[2]   Machine Technology Center Turku Ltd., 20520 Turku, Finland
\*   Correspondence: matti.manninen@lut.fi; Tel.: +358-40-574-2355

Received: 16 August 2017; Accepted: 4 September 2017; Published: 6 September 2017

**Abstract:** Laser drilling and laser engraving are common manufacturing processes that are found in many applications. With the continuous progress of additive manufacturing (3D printing), these processes can now be applied to the materials used in 3D printing. However, there is a lack of knowledge about how these new materials behave when processed or machined. In this study, sheets of 316L stainless steel produced by both the traditional cold rolling method and by powder bed fusion (PBF) were laser drilled by a nanosecond pulsed fiber laser. Results were then analyzed to find out whether there are measurable differences in laser processing parts that are produced by either PBF (3D printing) or traditional steel parts. Hole diameters, the widths of burn effects, material removal rates, and hole tapers were measured and compared. Additionally, differences in microstructures of the samples were also analyzed and compared. Results show negligible differences in terms of material processing efficiency. The only significant differences were that the PBF sample had a wider burn effect, and had some defects in the microstructure that were more closely analyzed. The defects were found to be shallow recesses in the material. Some of the defects were deep within the material, at the end and start points of the laser lines, and some were close to the surfaces of the sample.

**Keywords:** laser processing; laser drilling; laser engraving; additive manufacturing; 3D printing; powder bed fusion; stainless steel

---

## 1. Introduction

Laser drilling is a process where material is removed by a focused laser beam to produce usually a hole with a circular cross-section. Schulz et al. extensively covered the basics of laser drilling in a review [1]. It is used for a variety of applications, e.g., the production of cooling channels and lubrication holes for turbine components and engine parts, the fabrication of orifices for nozzles and controlled leaks, microvia hole drilling in circuit interconnects, and biomedical applications such as needles, catheters, and sensors [2,3]. Also, other electrical components such as hard disks, displays, computer peripherals, and telecommunication devices utilize laser drilling on a micro scale [3]. The four most used processes for laser drilling are single pulse, percussion (using multiple pulses in the same spot), trepanning (moving the laser beam along the circumference of the hole), and helical drilling (trepanning while simultaneously adjusting the focal plane).

Laser drilling is a relatively well-understood process that has been studied for decades. There are certain aspects of drilling that have been of special interest. Challenges related to producing high precision and high aspect ratio holes have been investigated [4]. Aspect ratios greater than 50 have been produced with high quality, usually by employing helical or trepanning drilling [5–7]. In addition to high precision and aspect ratios, high-speed drilling and micro drilling are utilized, especially in aerospace applications [8]. Burr formation and elimination are also important issues, studied e.g., by Duan et al. [9]. Their investigations showed which laser parameters cause burr formation and how to minimize it. Ghoreishi et al. investigated the effect of laser parameters on hole taper and circularity in percussion drilling of steels [10]. Goyal and Dubey investigated similar hole characteristics and drilling optimization in a case of titanium alloy material [11]. Their studies provide guidance on how to optimize processing parameters to produce smaller holes with less tapering, and better circularity. Hole quality properties, including circularity and taper, are also important because of the need to control the shape of the drilled hole [12]. Jahns et al. developed a new type of optical systems that enable control of the hole taper to produce either positive or negative taper [13,14]. Some effort has been expended to combine electro-discharge machining with laser drilling to produce high quality holes without the recast layer and heat affected zones, which can be problematic in laser drilling [15–17].

More specialized studies include Hidai et al.'s examination of curved drilling using inner hole reflections [18]. They were able to drill a 50-µm diameter hole with a curvature of about 3°, beginning at a depth of 400–600 µm. On a more fundamental level, Arrizubieta et al. investigated the hole characterization and formation [19]. They developed a numerical model to simulate the drilled hole geometry for AISI (American Iron and Steel Institute) 304 stainless steel. Jackson and O'Neill investigated the drilling of tool steels by using a nanosecond diode-pumped Nd:YAG (neodymium doped yttrium-aluminum-garnet) laser [20]. With a laser wavelength of 1064 nm, the maximum material removal rate obtained was about 1 µm per pulse. Voisey et al. studied single pulse drilling with a millisecond-pulsed Nd:YAG laser [21]. They determined particle size distribution, the angle of trajectory, molten layer thickness, and the temporal variation of melt ejection.

Drilling depth estimation can be difficult unless a vast experimental data bank is available. Ho et al. monitored the plasma plume from the side, and estimated the depth online through image acquisition and analysis [22]. Their results indicate that the cumulative plasma size correlates with the drilling depth, with a high degree of confidence.

Overall, there seems to be a reasonable amount of data about the principles and mechanisms of laser drilling, with many studies focusing on the effects of processing parameters on the quality and speed-related properties of the drilled hole. However, there are no studies conducted specifically on drilling laser additively manufactured materials. This is understandable, since the technology is still new and the differences between metallic materials produced conventionally and from melted powder seem to be quite small. This study aims to find out whether there are any significant differences between laser drilling a one-millimeter thick cold rolled AISI 316L stainless steel and a similar sheet produced by selective laser melting—one of the techniques of additive manufacturing. The results provided will support more fundamental research into the differences between conventionally produced and additively manufactured steel parts, as well as help the industry utilize laser engraving or drilling technology to better optimize their laser parameters.

In this study, the laser drilling is performed by engraving a circular hole through the material, because that method works best for the laser used in the study. While this is an unusual process, in this case the specific method of drilling is not important as the study is about the differences in processing two different materials. Laser engraving shares many similarities with laser drilling, because they are both processes that essentially require the removal of material until a desired shape is achieved. The major differences are that in engraving, the material is not usually fully penetrated, and the desired shapes are more complex.

## 2. Materials and Methods

The laser source used in the engraving experiments was a pulsed ytterbium fiber laser manufactured by IPG Photonics, with 20 W of maximum average power, typical beam quality $M^2$ value of 1.5, maximum pulse energy of 1 mJ, a variable pulse repetition rate from 1.6 to 1000 kHz, and a changeable pulse length from 4 ns to 200 ns in 8 different waveforms. Scan head optics was used where the scan head was Scanlab's Hurryscan 14 II with an f100 telecentric lens. The laser beam focal point diameter was ~40 μm with a Rayleigh length of 0.39 mm and a near Gaussian beam power distribution, as measured by Primes MicroSpotMonitor monitoring tool (see Figure 1b). A schematic drawing of the test setup can be seen in Figure 1a.

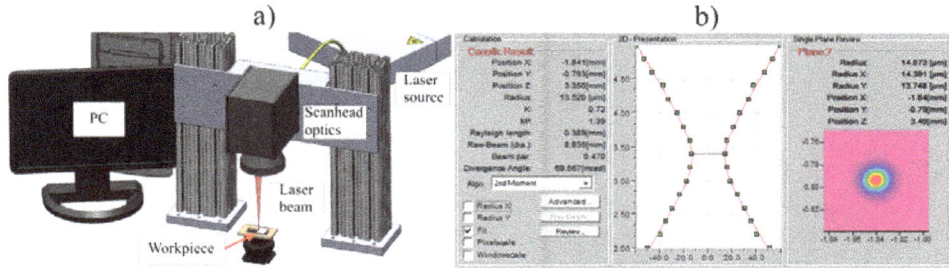

**Figure 1.** (**a**) A schematic illustration of the experimental setup; (**b**) Tested optical properties of the laser beam.

## 2.1. Test Sample Materials and Manufacture of PBF Sheet

Two sets of laser drilling experiments were conducted; one for a cold rolled 316L steel sheet and one for a powder bed fusion (PBF)-produced steel sheet. Both steel sheets were of the same size, $20 \times 30 \times 1$ mm. The compositions for both steels are shown in Table 1; the values were found in literature and measured values are given. The compositions were measured by Hitachi scanning electron microscope (model number SU3500) from the surface of the samples.

**Table 1.** Material compositions by mass percentages for the experimental samples.

| Material | C | Cr | Ni | Mo | Mn | Cu | Si | P | S | N | Fe |
|---|---|---|---|---|---|---|---|---|---|---|---|
| CR1 [1] | <0.03 | 17.2 | 10.1 | 2.1 | <2 | <0.5 | <0.75 | <0.045 | <0.03 | <0.1 | bal. |
| CR2 [2] | - | 18.4 | 9.5 | 2.3 | 2.1 | - | 0.6 | - | - | - | 66.7 |
| PBF1 [3] | <0.03 | 17–19 | 13–15 | 2.25–3 | <2 | <0.5 | <0.75 | <0.025 | <0.01 | <0.1 | bal. |
| PBF2 [4] | - | 18.4 | 11.4 | 2.4 | 1.3 | - | 0.7 | - | - | - | 65.5 |

[1] Cold rolled 316L, values from literature. [2] Cold rolled 316L, measured values. [3] PBF 316L, values from literature. [4] PBF 316L, measured values (PBF: powder bed fusion).

The PBF sheets were manufactured using the selective laser melting (SLM) technique [23] with a modified research PBF system, similar to EOSINT M270, manufactured by EOS, with a maximum build volume of $245 \times 245 \times 215$ mm. The system uses a 200 W ytterbium fiber laser to selectively melt powder layer by layer. The focal point diameter of the laser beam was 100 μm. Manufacture of the sheet was executed as one build. Nitrogen was used as the shielding gas in the process chamber to avoid oxidation of the parts and to reduce the oxygen content to below 0.8%. Powder was deposited by a recoater blade that moved in one direction from the powder storage platform, onto the building platform, and finally depositing the excess powder onto the third platform. All of the platforms moved vertically as necessary to build the sheet. For laser beam scanning, the 'island' scan strategy, as studied by Carter et al. [24], was used. The manufacturing parameters were: laser power 200 W, laser scan velocity 1000 mm/s, layer thickness 20 μm, and hatch distance 0.1 mm. After the manufacture, the sample was sawed off the building platform. No other post-processing was done. Figure 2 shows the orientation of the sample from the top-down perspective during the manufacture.

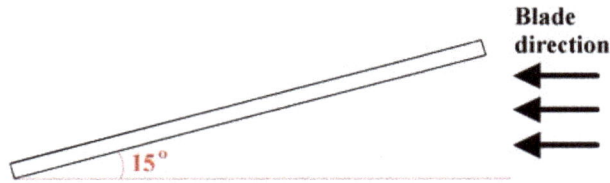

**Figure 2.** Top-down illustration of the orientation of the powder bed fusion (PBF) sheet during manufacture; thickness: 1 mm, length: 30 mm.

*2.2. Differences in Physical Properties*

Analyzing the differences between the physical properties of the two materials would be beneficial for understanding the possible reasons for processing them in different ways. However, while there is plenty of information available on traditionally produced 316L stainless steel, there is no information readily available on the thermal conductivity, thermal expansion coefficient, heat capacity, or many other properties of 3D printed stainless steels. The mechanical properties, such as tensile and yield strengths and Young's modulus, are available for both materials. Some typical physical properties of these materials are given in Table 2. For some additional insight into the differences, according to a data sheet published by 3D Alchemy [25], the thermal conductivity for 1.4542 (martensitic, AISI 630) stainless steel is 13 W/m °C as built, whereas for conventionally produced sheets it is 14 W/m °C at room temperature [26]. This indicates that the cold or hot rolled material might have a slightly higher thermal conductivity than the 3D printed parts. The 3D printed stainless steels seem to also be a little harder. Studies have indicated that the relative density of 3D printed parts is at around 99.5% [27,28].

**Table 2.** Comparison of physical properties between 3D-printed [29] and traditional 316L [30] stainless steel.

| Material | Ultimate Tensile Strength, MPa | Hardness, Rockwell B | Density, g/cm$^3$ | Coef. of Thermal Exp., °C$^{-1}$ | Thermal Conductivity, W/(m °C) | Heat Capacity, J/(g °C) |
|---|---|---|---|---|---|---|
| 316L | 586 | 81 | 7.9 | $1.602 \times 10^{-5}$ | 15.1 | 0.502 |
| 3D printed 316L | 640 ± 50 (horizontal) 540 ± 55 (vertical) | 85 | min. 7.9 | - | - | - |

In addition to the properties listed in Table 2, residual stresses within the material are different. The cold rolled sheet was annealed to remove external stresses, but the PBF sheet was used in the experiments as built. According to Wu et al., the PBF-produced 316L sheets typically have high compressive stress in the center, and tensile stress near surfaces [31]. Still, considering that the properties of traditional and 3D printed materials seem to be very close to each other, it can be expected that the results in laser engraving should differ by a few percentage points at most, despite how rough the surface of a 3D printed part is.

*2.3. Drilling Procedure and Parameters*

Seven holes were drilled on each plate, near the edge. Different waveforms and other laser parameters were used for each of the seven holes. The parameters are listed, and different waveforms shown in Table 3. Maximum pulse repetition rates (PRR) corresponding to the pulse length were used. Laser beam speed was kept at 2000 mm/s, except for those pulse lengths where the PRR was so low that the speed had to be lowered to keep the minimum pulse overlap. Minimum pulse overlap was 50% to ensure even material removal. Waveforms are shown in Table 3, where the X-axis is time, and the Y-axis is radiation intensity, in order to illustrate the temporal shapes of the laser pulses.

**Table 3.** Used laser parameters and waveforms for the experiments. PRR: pulse repetition rates.

| # | Pulse Length, ns | Pulse Energy, mJ | PRR, kHz | Pulse Overlap, % | Beam Speed, mm/s | Waveform |
|---|---|---|---|---|---|---|
| 1 | 8 | 0.100 | 200 | 75 | 2000 | |
| 2 | 14 | 0.160 | 125 | 60 | 2000 | |
| 3 | 20 | 0.190 | 105 | 52 | 2000 | |
| 4 | 30 | 0.235 | 85 | 50 | 1700 | |
| 5 | 50 | 0.333 | 60 | 50 | 1200 | |
| 6 | 100 | 0.500 | 40 | 50 | 800 | |
| 7 | 200 | 1.000 | 20 | 50 | 400 | |

Drilling was not performed by any of the common drilling methods such as percussion or trepanning techniques, but rather by engraving, in order to find out whether it can be a competitive method for producing holes. The engraving was performed by moving the laser beam in the pattern shown in Figure 3a. During drilling, the work piece was moved higher so that the focal point was always near the surface, until a through hole was achieved. The focal plane was adjusted up to 100 times, depending on the laser parameters. The drilling was stopped while adjusting the focus, which allowed the work piece to cool down between the cycles. After full penetration, up to 10 additional drilling cycles were repeated at the lowest focal point to ensure as clean an exit hole as possible. The holes were drilled near the edge (as seen in Figure 3b) to make it easy to produce micrographs from the cross-sections of the holes by grinding the excess material from the edge.

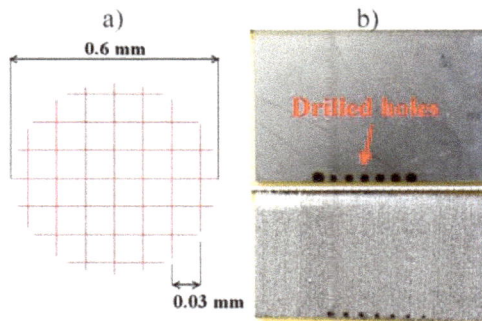

**Figure 3.** (**a**) Hatch pattern illustration for the engraving, not in scale. Red lines denote laser beam movement; (**b**) Photographs of the drilled sheets, drilled holes were 2 mm apart. Cold rolled sheet on top, PBF sheet on bottom.

### 2.4. Measured Properties

Material removal rates, entry and exit hole diameters, hole tapers, and the widths of the burning effect (BE) on the surface, near the entry and exit holes, were measured and compared. Additionally, some microstructural analysis was carried out. Figure 4 shows where the hole diameter and BE measurements were taken, with example images taken from the drilled cold rolled sheet. For the purpose of microstructural analysis, the two sample sheets were ground and polished as close to the middle line as possible to show the cross-sections of the holes. Polished samples were then etched to show the microstructure of the material. A light microscope and a scanning electron microscope were used to take magnified images of the hole profiles.

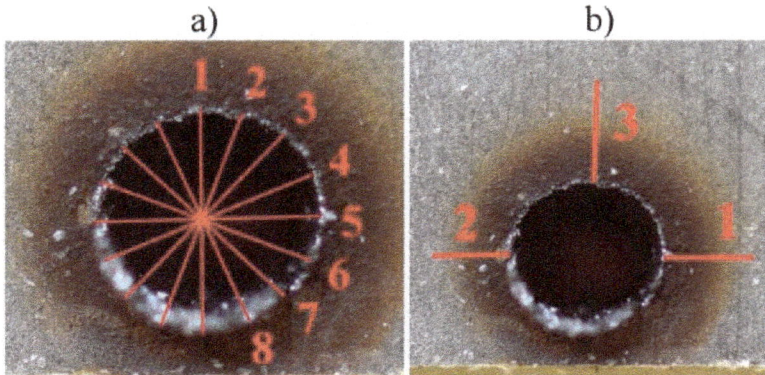

**Figure 4.** Images showing where the measurements were taken. (**a**) Eight measurements for the hole diameter; (**b**) Three measurements for the burning effect (BE) width.

Eight measurements were taken of each drilled hole (see Figure 4a), and the average value was used for the hole diameter calculations. Three measurements were taken of the BE (see Figure 4b) where the width of the BE was determined visually from the magnified images, and the average value was used. It was reasonably clear from the magnified images where the base material began to change color to yellow, but since the measuring process was done manually, there is likely some error present. The amount of possible error was estimated by measuring the distance from 'where the color of base material has clearly changed' to 'where the color of base material has clearly not changed'. The error estimation was done for each BE measurement. Entry and exit holes were measured separately for both the hole diameter and BE. Hole tapers in degrees were calculated by a simple trigonometric equation:

$$\text{Hole taper} = (180/\pi) \tan^{-1} (r_1 - r_2)/h \tag{1}$$

where $r_1$ and $r_2$ are the radii of the entry and exit holes, and h is the thickness of the material.

Material removal rate was calculated by the removed volume and measured drilling time. Since the drilled holes were shaped almost exactly like circular truncated cones, the volume was easy to calculate from knowing the entry and exit hole radii, and the thickness of the material. Exact material thicknesses were measured from the micrographs and used in the calculations. Since the cold rolled steel sheet was found to be slightly thicker, the exit hole diameters for the cold rolled samples were measured at a depth corresponding to the thickness of the PBF sample. Volume was then divided by the drilling time to get the material removal rate (MRR) in cubic millimeters per minute.

Scanning electron microscope (SEM) images of both the PBF and cold-rolled steel samples were taken to examine any significant differences in the material microstructure. A Hitachi SU3500 scanning electron microscope was used for the imaging.

## 3. Results and Discussion

The main results of the study are compiled into the following five sections. The study focuses on comparing the results to find differences between laser engraving of cold rolled steel and PBF steel. The individual quality of the holes is of less importance. Magnified images of the entry and exit holes, and the micrographs of the hole profiles, are shown in Figure 5. Images for the cold-rolled steel sheet are shown on the left, and for the PBF sheet on the right.

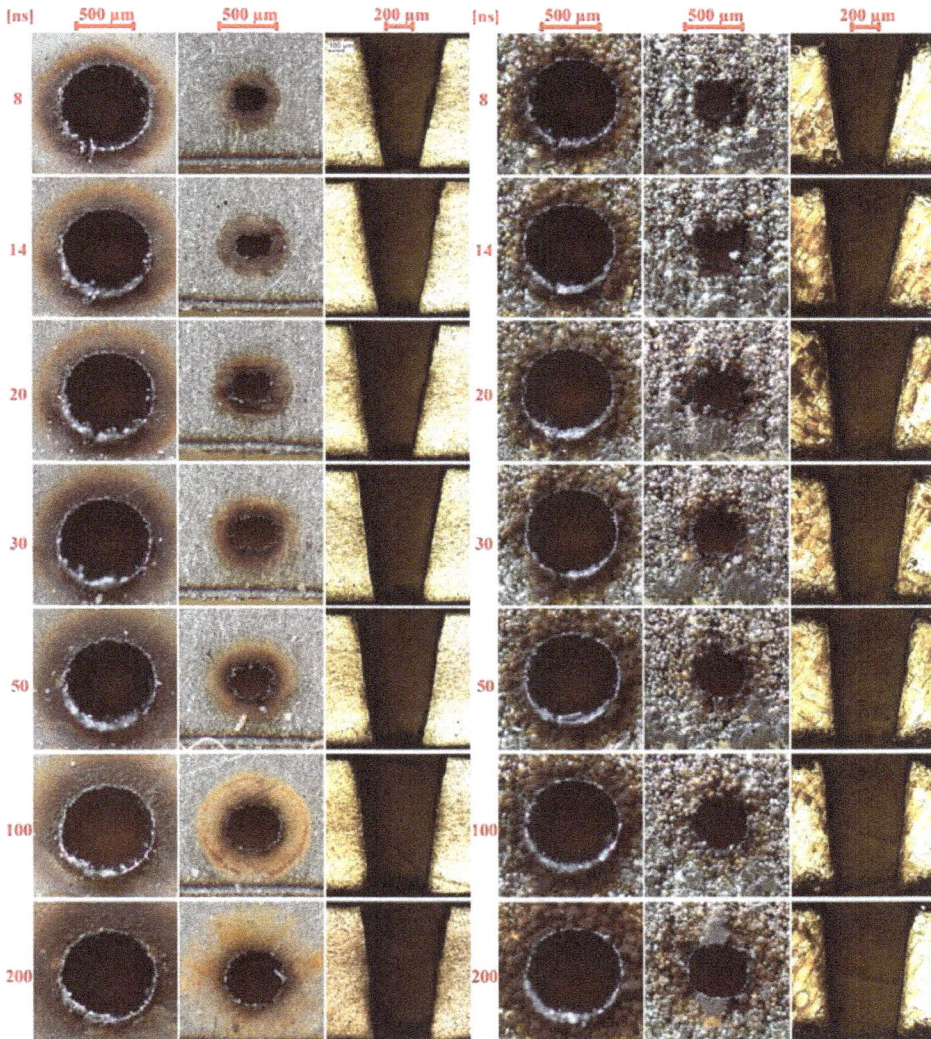

**Figure 5.** Magnified images of the entry and exit holes, and hole profiles, respectively, for all drilled holes. Cold-rolled sheet images are on the left, and images from the PBF sheet are on the right.

### 3.1. Hole Diameters

Hole diameters for the entry and exit holes can be seen in Figure 6. The target diameter for the holes were 0.6 mm, so it can be seen that the entry holes are a bit larger and the exit holes are much smaller than the target diameter. Too large entry holes can usually be fixed by using a large enough beam offset with the control program. However, even with corrections, it is difficult to prevent entry hole erosion due to the upwards melt flow, as discussed by Li et al. [32]. In practice, the entry hole will always be slightly rounded in laser drilling, unless the material is directly vaporized by a high enough power density. The hole diameter clearly increases with increasing pulse length. This can be attributed to the higher pulse energy and longer interaction with the material, which leads to more molten material being ejected by vapor pressure. When the pulse length is longer than 50 ns pulses, the hole diameters seem to increase nearly linearly with the increase in pulse length. When the pulse length is shortened to less than 50 ns, the hole diameters start to decrease rapidly. It is commonly known that when the pulse length is reduced to the picosecond range, the laser-induced material removal starts to resemble so-called cold ablation, where material is removed with only minimal heat effect [33]. The decrease in hole diameter seen here is likely the beginning of a transition from a highly thermal process, where the primary mechanism of material removal is the ejection of molten material by vapor pressure, to material removal by direct evaporation.

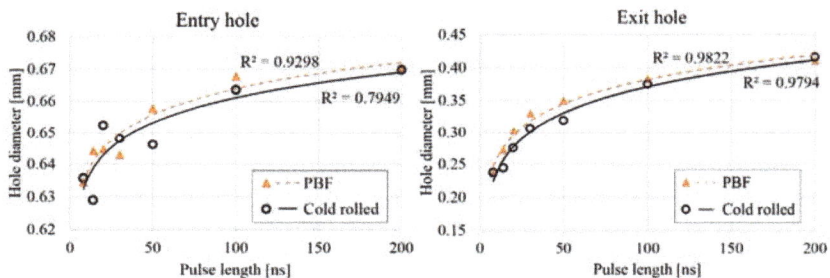

**Figure 6.** Measured entry and exit hole diameters for the laser drilled holes.

According to the results, there is at most a very slight difference between laser drilling cold rolled steel and PBF steel. On both sides, the differences are almost negligible. In four instances, the holes in cold rolled sheet are slightly larger, but overall the holes in PBF sheets are slightly larger.

### 3.2. Material Removal Rate

Material removal rates for all of the drilled holes are shown in Figure 7. The process becomes significantly faster with the increase in pulse length: the drilling is over six times faster with the longest pulse length compared to the shortest. The increase is fairly linear. Similar results have been published by von Allmen, who investigated the effect of pulse duration on drilling depth in the microsecond range, and Manninen et al., who investigated the same in the nanosecond range [34,35]. No significant difference can be noticed between the cold-rolled sheets and the PBF sheets. Generally, the MRR is about 0–10% higher when PBF sheets were drilled, except in the 200 ns case, when the cold-rolled sheet was drilled 4% faster.

Similar to the results in hole diameters, the significant increase in material removal rate following an increase in pulse length can be attributed to a longer interaction time between the laser pulse and the base material. As the laser pulse heats up and eventually melts a portion of the base material, the resulting melt pool will be ejected from the bottom of the drilled hole by vapor pressure. The longer the interaction time, the more material will be melted, and thus ejected. It is surprising how significant this effect is, given that the same average laser power was used for both 200 ns and 8 ns processing.

Compared with drilling with 200 ns pulses, six times more energy had to be used when drilling with 8 ns pulses. The excess energy likely heated up the base material, provided the beam was not reflected. However, compared with longer pulses, the melt ejection phenomenon could not last as long, since the pulses were much shorter. Instead, the majority of the pulse energy likely went into heating up the base material instead of material removal. As the pulse length was increased, a higher portion of the pulse energy was expended in actual material removal once the material was molten. This problem seems to exist only in the range of tens of nanoseconds to hundreds of picoseconds, because at ultrashort pulse regime (picosecond to femtosecond), the power density is so high that the material is directly vaporized.

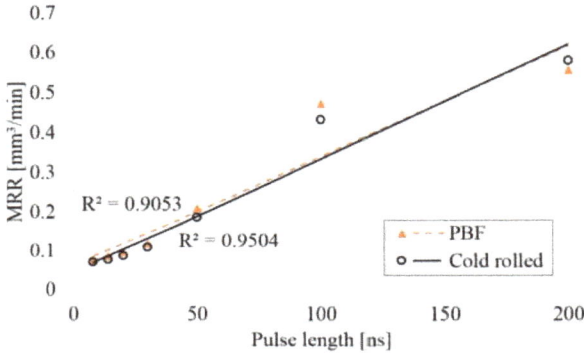

**Figure 7.** Material removal rates for holes laser drilled in PBF and cold rolled sheets.

*3.3. Burning Effect Width*

BE width for both the entry and exit holes can be seen in Figure 8. The BE width seems to increase nearly linearly in all cases, which indicates that the processing temperatures increase with the increase in pulse length. At the shortest pulse lengths, the BE width seems to start decreasing more rapidly. According to the heat temperature map published by the British Stainless Steel Association, the yellow to dark brown color seen in the samples indicate temperatures in the range of 57–673 K (30–400 °C) [36]. This agrees well with the previous findings published by the authors [34] where the processing temperature in engraving was measured with thermocouples.

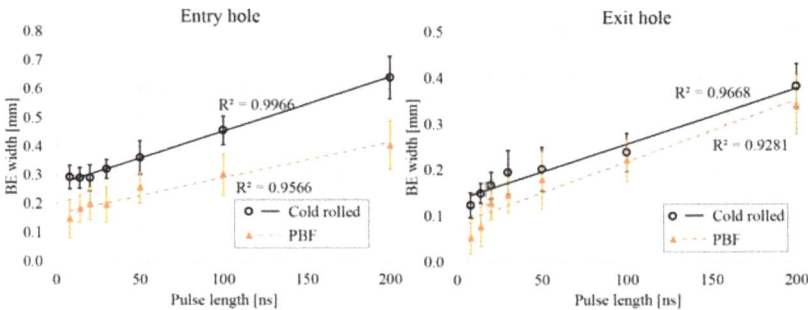

**Figure 8.** Measured BE widths for the entry and exit holes.

There are significant differences between the samples, even after accounting for the darker color and rough surface of the PBF sheets that made it more difficult to estimate where the color began to change. For the entry hole, there is on average almost a 0.25 mm difference in the width of the BE, which indicates higher temperatures on the surface of the cold-rolled sheet. It should be noted, however, that the surface of the PBF sheet is rough and consists of ball-like formations. The roughness of the surface can be seen from the magnified images taken of the surface of the PBF sheet and from the hole profile micrographs. The PBF sheets clearly have a much larger effective surface area. The increase in surface area should help dissipate heat faster, thus resulting in a smaller BE on the PBF sheet.

### 3.4. Hole Taper

Hole taper in degrees for all the drilled holes is shown in Figure 9. It can be seen that the taper decreases with increase in pulse length for both materials. No significant difference in hole taper between the two sheets can be recognized; in four cases, the holes drilled in the PBF sheet have higher tapering, and in three cases, the holes drilled in cold-rolled steel sheet have higher tapering. At its highest, the difference is 0.7 percentage units, but mostly it is much less than that.

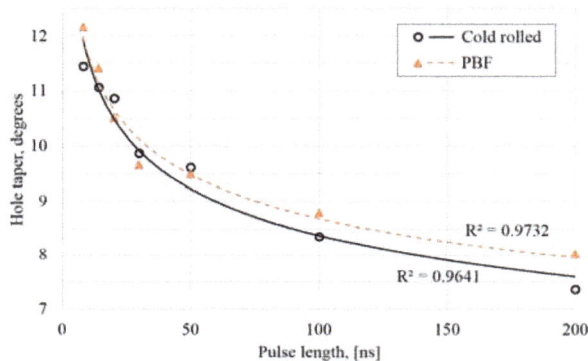

**Figure 9.** Measured hole tapers in degrees for all the drilled holes.

More interesting than the amount of taper is the shape of the tapering. Hole tapering in laser drilling has been reported many times, but it is almost always shaped like letter Y, i.e., the taper is not constant throughout. Instead, there is significant tapering close to the entry hole, after which the width of the hole stays mostly constant [32,37]. Drilling experiments in this study resulted in almost perfectly conical holes, due to the technique that was employed. Based on these results, it is difficult to say what exactly causes the taper angle to increase so rapidly with the decrease in pulse length. It can be concluded that holes with well-defined edges, but with high tapering, can be achieved with drilling by engraving without special optics to tilt the laser beam.

### 3.5. Evaluation of the Microstructures

The first micrographs of the samples were taken with a light microscope. Figure 10 illustrates the differences in microstructure between the cold-rolled and PBF sheets. The images were taken of the same surface as the hole profiles. Thus, the PBF sheet shows one melted layer, instead of a cross-section of multiple layers.

The cold-rolled sheet seen in Figure 10a has a typical austenitic stainless steel microstructure with a homogenous grain size and structure. No pores or other defects could be seen. The PBF sheet shows a much more irregular grain size pattern, but it still seems to be solid material with only few small pores. However, there are some circular patterns that can be seen as black lines in the PBF micrographs (examples circled in red in Figure 10b). After comparing them to the laser hatch pattern used in the production of the PBF sheet, it can be said that these are likely either the start or end points of laser lines. To find out what these dark lines look like at higher magnification, a scanning electron microscope was employed to analyze the topography of the surface. In Figure 11, a topographic image of a similar occurrence of a semicircle-like pattern can be seen.

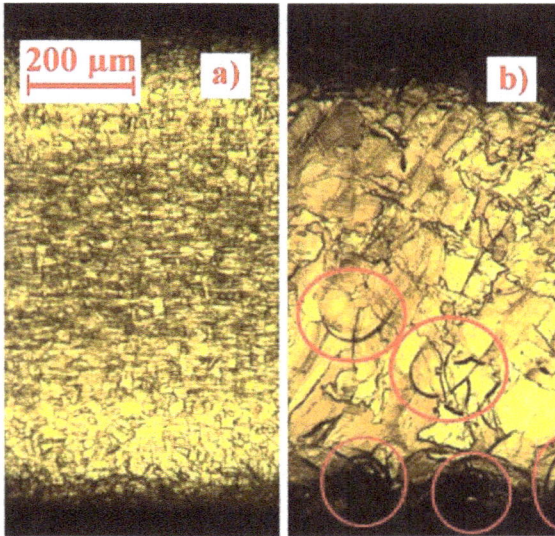

**Figure 10.** Magnified images of the samples showing the microstructure; (**a**) Cold-rolled steel sheet; (**b**) PBF produced sheet. Areas of interest circled in red.

From Figure 11, it can clearly be seen that the black lines in the PBF material are shallow recesses in the material. It can also be seen that they do not follow the grain boundaries. They are only about 10 μm deep and 10 μm wide and exist near the bottom and top surfaces of the material, and at the end and start points of the laser lines used in the hatch pattern when the sheet was produced. Overall, there seem to be too few of these defects, and they are not large enough to have a significant effect on the engraving process.

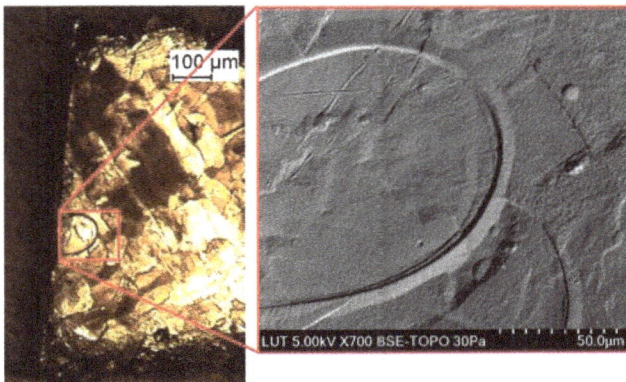

**Figure 11.** Scanning electron microscope (SEM) image showing the topography of an area with a suspected defect in the PBF sheet.

## 4. Conclusions

According to the results, the differences between laser engraving PBF and cold-rolled steel sheets seem to be negligible, or at most very small. The material removal rate, including drilling time and removed volume of material, was 0–10% higher for the PBF sheets. Generally, the holes drilled in PBF sheets were slightly bigger in diameter, and had a slightly higher tapering, up to 0.7 percentage points. The width of the BE on the material surface at the entry side was up to 0.24 mm smaller for the PBF sheets, which is a significant difference, since the drilled holes were only 0.6 mm in diameter. The difference in BE widths were attributed to the different surface topographies. From the magnified images of the PBF material surface, it can be easily seen that the surface of the material is very rough. The surface consists of small round particles with an average particle size of 32 μm (100 measurements), which is very close to the raw powder average particle size of 40 μm. The much higher effective surface area of PBF sheets likely dissipates heat more effectively, which results in a smaller BE. It should also help with the initial absorption of the laser beam, but in deep engraving this is not very significant.

Some defects were found when analyzing the microstructure of the PBF sheet. Semicircular black traces were seen at the start and end points of the laser lines, as well as near the surfaces of the PBF material. These were found to be shallow recesses in the material. It is unclear whether they were produced by the PBF-building process, or by the grinding and polishing when the samples were prepared for the micrographs. A likely explanation would be that there are deviations in energy input at the start and end points of the laser lines that causes detrimental changes in the material structure, or causes some of the powder material to not properly melt. The nearly loose material that was not fully fused was then removed when the sample was prepared for analysis. Near the surfaces of the material, these circular defects are not semicircular anymore, but rather full circles of shallow recesses. It seems as if a few powder particles were fused together, but were not fully fused to the surrounding material. Thus, two types of visible defects were noticed in the PBF material compared with the cold-rolled material: (1) powder particles near the surfaces of the object that are not fully fused to the surrounding material, and (2) semicircular defects at the end and/or start points of the laser lines. However, it can be concluded that it is very unlikely that these had a significant effect on the laser engraving process, since the recesses are so small and sparsely located.

**Acknowledgments:** This study was conducted in Finland as part of the research project "Micro- and millistructured reactors for catalytic oxidation reactions", abbreviated MICATOX. The financial support of The Academy of Finland and the participants of the project are gratefully acknowledged.

**Author Contributions:** Matti Manninen, Ville-Pekka Matilainen and Antti Salminen conceived and designed the experiments; Matti Manninen performed the experiments; Matti Manninen and Marika Hirvimäki analyzed the data and wrote the paper.

**Conflicts of Interest:** The authors declare no conflict of interest. The founding sponsors had no role in the design of the study; in the collection, analyses, or interpretation of data; in the writing of the manuscript, and in the decision to publish the results.

## References

1.  Schulz, W.; Eppelt, U.; Poprawe, R. Review on laser drilling I. Fundamentals, modeling, and simulation. *J. Laser Appl.* **2013**, *24*, 1–17. [CrossRef]
2.  Rohde, H. Drilling of Metals. In *LIA Handbook of Laser Materials Processing*, 1st ed.; Ready, J.F., Farson, D.F., Eds.; Magnolia Publishing Inc.: Pineville, LA, USA, 2001; pp. 474–477, ISBN 0-941463-02-8.
3.  Gower, M. Industrial applications of laser micromachining. *Opt. Express* **2000**, *7*, 56–67. [CrossRef] [PubMed]
4.  Uchtmann, H.; He, C.; Gillner, A. High precision and high aspect ratio laser drilling—Challenges and Solutions. In Proceedings of the SPIE 9741, High-Power Laser Materials Processing: Lasers, Beam Delivery, Diagnostics, and Applications V, San Francisco, CA, USA, 13 February 2016; Volume 974106, pp. 1–12.
5.  Weck, A.; Crawford, T.H.R.; Wilkinson, D.S.; Haugen, H.K.; Preston, J.S. Laser drilling of high aspect ratio holes in copper with femtosecond, picosecond and nanosecond pulses. *Appl. Phys. A* **2008**, *90*, 537–543. [CrossRef]

6.  Patwa, R.; Herfurth, H.; Flaig, R.; Christophersen, M.; Philips, B.F. Laser Drilling for High Aspect Ratio Holes and a High Open Area Fraction for Space Applications. In Proceedings of the International Congress on Applications of Lasers & Electro-Optics, ICALEO, San Diego, CA, USA, 19–23 October 2014; Paper No. M1101; pp. 204–210.

7.  Fornaroli, C.; Holtkamp, J.; Gillner, A. Laser-beam helical drilling of high quality micro holes. *Phys. Procedia* **2013**, *41*, 661–669. [CrossRef]

8.  Stephen, A.; Schrauf, G.; Mehrafsun, S.; Vollertsen, F. High speed laser micro drilling for aerospace applications. *Procedia CIRP* **2014**, *24*, 130–133. [CrossRef]

9.  Duan, W.; Wang, K.; Dong, X. Experimental characterizations of burr deposition in Nd:YAG laser drilling: A parametric study. *Int. J. Adv. Des. Manuf. Technol.* **2015**, *76*, 1529–1542. [CrossRef]

10. Ghoreishi, M.; Low, D.K.Y.; Li, L. Comparative statistical analysis of hole taper and circularity in laser percussion drilling. *Int. J. Mach. Tools Manuf.* **2002**, *42*, 985–995. [CrossRef]

11. Goyal, R.; Dubey, A.K. Modeling and optimization of geometrical characteristics in laser trepan drilling of titanium alloy. *J. Mech. Sci. Technol.* **2016**, *30*, 1281–1293. [CrossRef]

12. Begic-Hajdarevic, D.; Bijelonja, I. Experimental and numerical Investigation of Temperature Distribution and Hole Geometry during Laser Drilling Process. *Procedia Eng.* **2014**, *100*, 384–393. [CrossRef]

13. Ashkenasi, D.; Kaszemeikat, T.; Mueller, N.; Dietrich, R.; Eichler, H.J.; Illing, G. Laser Trepanning for Industrial Applications. *Phys. Procedia* **2011**, *12*, 323–331. [CrossRef]

14. Jahns, D.; Kaszemeikat, T.; Mueller, N.; Ashkenasi, D.; Dietrich, R.; Eichler, H.J. Laser trepanning of stainless steel. *Phys. Procedia* **2013**, *41*, 630–635. [CrossRef]

15. Li, L.; Diver, C.; Atkinson, J.; Giedl-Wagner, R.; Helml, H.J. Sequential Laser and EDM Micro-drilling for Next Generation Fuel Injection Nozzle Manufacture. *CIRP Ann.-Manuf. Technol.* **2006**, *55*, 179–182. [CrossRef]

16. Uhlmann, E.; Oberschmidt, D.; Langmack, M. Complex bore holes fabricated by combined Helical Laser Drilling and Micro Electrical Discharge Machining. In Proceedings of the 28th Annual Meeting of the American Society for Precision Engineering, ASPE 2013, Saint Paul, MN, USA, 20–25 October 2013; pp. 195–199.

17. Luft, A.; Franz, U.; Emsermann, A.; Kaspar, J. A study of thermal and mechanical effects on materials induced by pulsed laser drilling. *Appl. Phys. A* **1996**, *63*, 93–101. [CrossRef]

18. Hidai, H.; Kuroki, Y.; Matsusaka, S.; Chiba, A.N.; Morita, N. Curved drilling via inner hole laser reflection. *Precis. Eng.* **2016**, *46*, 96–103. [CrossRef]

19. Arrizubieta, I.; Lamikiz, A.; Martínez, S.; Ukar, E.; Tabernero, I.; Girot, F. Internal characterization and the hole formation mechanism in the laser percussion drilling process. *Int. J. Mach. Tools Manuf.* **2013**, *75*, 55–62. [CrossRef]

20. Jackson, M.J.; O'Neill, W. Laser micro-drilling of tool steel using Nd:YAG lasers. *J. Mater. Process. Technol.* **2003**, *142*, 517–525. [CrossRef]

21. Voisey, K.T.; Kudesia, S.S.; Rodden, W.S.O.; Hand, D.P.; Jones, J.D.C.; Clyne, T.W. Melt ejection during laser drilling of metals. *Mater. Sci. Eng.* **2003**, *A356*, 414–424. [CrossRef]

22. Ho, C.; He, J.; Liao, T. On-Line Estimation of Laser-Drilled Hole Depth Using a Machine Vision Method. *Sensors* **2012**, *12*, 10148–10162. [CrossRef] [PubMed]

23. Casalino, G.; Campanelli, S.L.; Contuzzi, N.; Ludovico, A.D. Experimental investigation and statistical optimisation of the selective laser melting process of a maraging steel. *Opt. Laser Technol.* **2015**, *65*, 151–158. [CrossRef]

24. Carter, L.N.; Martin, C.; Withers, P.J.; Attallah, M.M. The influence of the laser scan strategy on grain structure and cracking behavior in SLM powder-bed fabricated nickel superalloy. *J. Alloys Compd.* **2014**, *615*, 338–347. [CrossRef]

25. Stainless Steel—1.4542 (Stainless 17.4) Datasheet (Default) by 3D Alchemy. Available online: https://www.3d-alchemy.co.uk/assets/datasheets/3d-Alchemy-Stainless%20Steel%201-4542%20-%2001_13.pdf (accessed on 25 August 2017).

26. 1.4542 Steel Datasheet by Lucefin. Available online: http://www.lucefin.com/wp-content/files_mf/1.4542pha63062.pdf (accessed on 25 August 2017).

27. Cherry, J.A.; Davies, H.M.; Mehmood, S.; Lavery, N.P.; Brown, S.G.R.; Sienz, J. Investigation into the effect of process parameters on microstructural and physical properties of 316L stainless steel parts by selective laser melting. *Int. J. Adv. Manuf. Technol.* **2014**, *76*, 869–879. [CrossRef]

28. Spierings, A.B.; Levy, G. Comparison of density of stainless steel 316L parts produced with selective laser melting using different powder grades. In Proceedings of the Twentieth Annual International Solid Freeform Fabrication (SFF) Symposium, Austin, TX, USA, 3–5 August 2009; pp. 342–353.

29. Think3d Web Page. Available online: https://www.additively.com/en/material/from/eos/eos-stainless-steel-316l (accessed on 25 August 2017, requires login).

30. Outokumpu Data Sheet for 316 and 316L. Available online: http://www.outokumpu.com/SiteCollectionDocuments/Datasheet-316-316L-imperial-hpsa-outokumpu-en-americas.pdf (accessed on 25 August 2017).

31. Wu, A.S.; Brown, D.W.; Kumar, M.; Gallegos, G.F.; King, W.E. An Experimental investigation into Additive Manufacturing-Induced Residual Stresses in 316L Stainless Steel. *Metall. Mater. Trans. B* **2014**, *45*, 6260–6270. [CrossRef]

32. Li, L.; Low, D.K.Y.; Ghoreshi, M.; Crookall, J.R. Hole Taper Characterisation and Control in Laser Percussion Drilling. *CIRP Ann.-Manuf. Technol.* **2002**, *51*, 153–156. [CrossRef]

33. Chickov, B.N.; Momma, C.; Nolte, S.; von Alvensleben, F.; Tünnermann, A. Femtosecond, picosecond and nanosecond laser ablation of solids. *Appl. Phys. A* **1996**, *63*, 109–115. [CrossRef]

34. Von Allmen, M. Laser drilling velocity in metals. *J. Appl. Phys.* **1976**, *47*, 5460–5463. [CrossRef]

35. Manninen, M.; Hirvimäki, M.; Poutiainen, I.; Salminen, A. Effect of Pulse Length on Engraving Efficiency in Nanosecond Pulsed Laser Engraving of Stainless Steel. *Metall. Mater. Trans. B* **2015**, *46*, 2129–2136. [CrossRef]

36. British Stainless Steel Association. Available online: http://www.bssa.org.uk/topics.php?article=140 (accessed on 15 April 2017).

37. Bandyopadhyay, S.; Sarin Sundar, J.K.; Sundararajan, G.; Joshi, S.V. Geometrical features and metallurgical characteristics of Nd:YAG laser drilled holes in thick IN718 and Ti-6Al-4V sheets. *J. Mater. Process. Technol.* **2002**, *127*, 83–95. [CrossRef]

*applied sciences*

MDPI

*Article*

# Optimization of AZ91D Process and Corrosion Resistance Using Wire Arc Additive Manufacturing

**Seungkyu Han, Matthew Zielewski** [ID]**, David Martinez Holguin, Monica Michel Parra and Namsoo Kim *** [ID]

Department of Metallurgical, Materials and Biomedical Engineering, The University of Texas at El Paso, El Paso, TX 79968, USA; shan5@miners.utep.edu (S.H.); mzielewski@miners.utep.edu (M.Z.); damartinezholguin@miners.utep.edu (D.M.H.); mmlugo2@utep.edu (M.M.P.)
* Correspondence: nkim@utep.edu; Tel.: +1-915-747-7996; Fax: +1-915-747-8036

Received: 18 July 2018; Accepted: 31 July 2018; Published: 6 August 2018

**Abstract:** Progress on Additive Manufacturing (AM) techniques focusing on ceramics and polymers evolves, as metals continue to be a challenging material to manipulate when fabricating products. Current methods, such as Selective Laser Sintering (SLS) and Electron Beam Melting (EBM), face many intrinsic limitations due to the nature of their processes. Material selection, elevated cost, and low deposition rates are some of the barriers to consider when one of these methods is to be used for the fabrication of engineering products. The research presented demonstrates the use of a Wire and Arc Additive Manufacturing (WAAM) system for the creation of metallic specimens. This project explored the feasibility of fabricating elements made from magnesium alloys with the potential to be used in biomedical applications. It is known that the elastic modulus of magnesium closely approximates that of natural bone than other metals. Thus, stress shielding phenomena can be reduced. Furthermore, the decomposition of magnesium shows no harm inside the human body since it is an essential element in the body and its decomposition products can be easily excreted through the urine. By alloying magnesium with aluminum and zinc, or rare earths such as yttrium, neodymium, cerium, and dysprosium, the structural integrity of specimens inside the human body can be assured. However, the in vivo corrosion rates of these products can be accelerated by the presence of impurities, voids, or segregation created during the manufacturing process. Fast corrosion rates would produce improper healing, which, in turn, involve subsequent surgical intervention. However, in this study, it has been proven that magnesium alloy AZ91D produced by WAAM has higher corrosion resistance than the cast AZ91D. Due to its structure, which has porosity or cracking only at the surface of the individual printed lines, the central sections present a void-less structure composed by an HCP magnesium matrix and a high density of well dispersed aluminum-zinc rich precipitates. Also, specimens created under different conditions have been analyzed in the macroscale and microscale to determine the parameters that yield the best visual and microstructural results.

**Keywords:** additive manufacturing; WAAM; corrosion; AZ91D

## 1. Introduction

The application of ceramics, polymers, and polymer-matrix composites for bone healing procedures has been continuously increasing [1,2], especially in metals such as stainless steel, titanium and cobalt-based alloys, which are the most widely used composites to heal or replace damaged bones [3]. However, devices with these composites have a higher strength moduli. This may cause stress shielding to occur around the area where the device is implanted. The reduction in stress leads to a decrease in bone density on the areas close to the metallic component, which may lead to a deficient healing process [4]. Furthermore, limitations arise from the possible liberation of harmful metallic ions inside the organism due to corrosion mechanisms, as well as debris formation because of wear and tear that lead to inflammation in the areas near the biomedical device [5].

For alternative bone healing purposes, magnesium has been studied due to its mechanical properties and degradability inside the human body. As shown in Table 1, the elastic modulus and yield strength of magnesium are closer to those of human bone, reducing the stress shielding phenomena [6]. As the implant decomposes, magnesium elements will progressively pass through the human body upon completing the assigned task (bone healing) until no trace of the original element is present inside the organism [7]. The body requires magnesium ions for both metabolic reactions and biological mechanisms. Typically, a 70 kg person would have approximately 35 g of magnesium [8]. Excess magnesium ions are excreted through urine with no harm on the metabolic system [9].

**Table 1.** Mechanical properties of steels and bone. Reproduced from [6]. Copyright 2006, Elsevier.

|  | Density | Elastic Modulus (GPa) | Compressive Strength (MPa) | Fracture Toughness (MPa/m$^2$) |
|---|---|---|---|---|
| Natural Bone | 1.8–2.1 | 3–20 | 130–180 | 3–6 |
| Ti alloy | 4.4–4.5 | 110–170 | 758–1117 | 55–115 |
| Co-Cr Alloy | 8.3–9.2 | 230 | 450–1000 | N/A |
| Stainless Steel | 7.9–8.1 | 189–205 | 170–310 | 50–200 |
| Magnesium | 1.74–2.0 | 41–45 | 64–100 | 15–40 |

However, one of the main shortcomings of magnesium alloy implants is fast degradation via corrosion [10–12]. A new manufacturing process to produce magnesium implants can assist in corrosion resistance. This study demonstrates that magnesium produced by Wire and Arc Additive Manufacturing (WAAM) technique creates impurity-free and void-free structures that will increase the corrosion resistance [12]. AZ91D was selected since it is a low-cost, readily available alloy, which mainly consists of aluminum, zinc, and magnesium as referenced in Table 2 [13]. To determine the microstructure and corrosion performance of AZ91D produced by WAAM, it was studied with electrochemical impedance spectroscopy (EIS), scanning electron microscopy (SEM), X-ray diffraction (XRD), energy dispersive spectroscopy (EDS), and optical microscopy.

**Table 2.** Chemical composition of magnesium alloy AZ91D. Reproduced from [14]. Copyright 1998, Elsevier.

| Element | Wt % |
|---|---|
| Al | 8.5–9.5 |
| Zn | 0.45–0.9 |
| Si | 0.05 max |
| Mn | 0.17 max |
| Cu | 0.015 max |
| Fe | 0.004 max |
| Ni | 0.001 max |
| Other | 0.01 max |
| Mg | Balance |

## 2. Materials and Methods

### 2.1. WAAM (Wire and Arc Additive Manufacturing)

SLS and EBM techniques rely either on a high energy laser or an electron beam for the creation of 3D structures. Both cases required large amounts of energy for the creation of these heat sources. The creation of a localized atmosphere is also required for powder bed and powder feed systems. In laser sintering techniques the temperature of the chamber must be increased to almost the melting temperature of the material which is being printed. To achieve these temperatures, a large consumption of energy is required. In the case of EBM, the necessity of a high vacuum also affects the required energy input [14]. Wire and Arc Additive Manufacturing (WAAM) combines an electric arc (heat source) and a metal wire as feedstock to create 3D models in a bottom-up approach (layer upon layer) [15]. Due to its simplified design and the feedstock, WAAM presents many advantages over powder bed and powder feed systems. WAAM requires less energy consumption when compared to powder bed methods because there is no need for a high-powered energy source to heat the material to melting temperature. In contrast to EBM techniques that use accelerating voltages from 30 to 60 kV [15], Plasma Arc Welding (PAW) systems can be used on 110 V configurations. Furthermore, powdered metal is made up of individual morphed spheres in different sizes. This dispersion of metal restricts the heat transfer and complicates the fusion across particle boundaries. Due to the lack of continuity in the structure, more heat is required to commence phase transformation [16–18]. Thus, less energy is necessary for melting wire metals, making it more energetically efficient [19].

### 2.2. Experimental Setup

The first step in the creation of the WAAM equipment was the development of a motion control system. For this purpose, aluminum frames were used to allow the fabrication of relatively large and heavy components without compromising the structural integrity of the equipment. The system is capable to move in three axes (*X*, *Y*, and *Z*). Two different frames were created. The first one is used to control the movement of the printing platform in the *X* and *Y* axes. This frame is placed horizontally, directly over the horizontal plane. The printing bed is belt-driven by two 5V stepper motors (one for each axis). Each axis slides with the use of four barrel-bearings over two steel rods (two bearings per rod), to ensure smooth movement. To hold the substrate in place, a clamping system made from aluminum was created. The *Z* axis is controled by an individual frame. A 12 V motor (used on a Lincoln Electric 140 c MIG welder (Lincoln Electric, Houston, TX, USA)) is used to feed wire. The system is controlled by a couple formed by an Arduino Mega 2650 board and a Ramps 1.4 board (JAMECO Electronics, Belmont, CA, USA). A modified version of the open source firmware Marlin (1.1, Ultimaker, Geldermalsen, NED) is used as operating environment (shown in Figure 1a).

A Pro-Fusion Dual Arc 82HFP (Elderfield & Hall, Knoxville, TN, USA) (see Figure 1b) welding machine was selected since it allows the precise control of the current intensity supplied to the system. In low current mode (from 0.1 to 20 Ampere) the parameter can be changed by increments of 0.1 Ampere. When working with higher currents (up to 80 Ampere) the steps become of 0.5 Ampere. The incorporation of "Pulsation Mode", in which the current is supplied in alternating on-off intervals, has been found convenient for the reduction of heat affected zones while working with stainless steel. The machine is also equipped with a "High Frequency" option, in which the machine sends high frequency pulses through the electrode until a grounded conductive surface is detected. Once a suitable surface is found, an electric arc is automatically started, and the electrode stops emitting high frequency pulses. This mode is highly convenient while printing since the arc will not always remain on. Two different heat sources have been used in this project.

The created WAAM systems rely on the use of a plasma arc for the melting and fusing of a metallic wire (filler) and a printing bed (substrate). The wire is being fed by 12 V motor through a reinforced conduction line and delivered to the molten metal pool created by the arc. The wire and substrate fuse together while the printing bed moves, creating a 3D line. The final build is affected by several factors, including printing bed speed, wire diameter, arc current density, the position and speed of the wire being fed through. Adjacent lines can be created by directly overlapping the line already printed and the printing path of the new line. Also, lines can be printed upon previous weld lines in order to create 3D structures. WAAM has already proved with our previous work (see Figure 1c,d) that it can produce 3D structured metal. Figure 1c shows 3D structured stainless steel stacked up with 24 lines. Figure 1d shows square shape of 3D printed stainless steel.

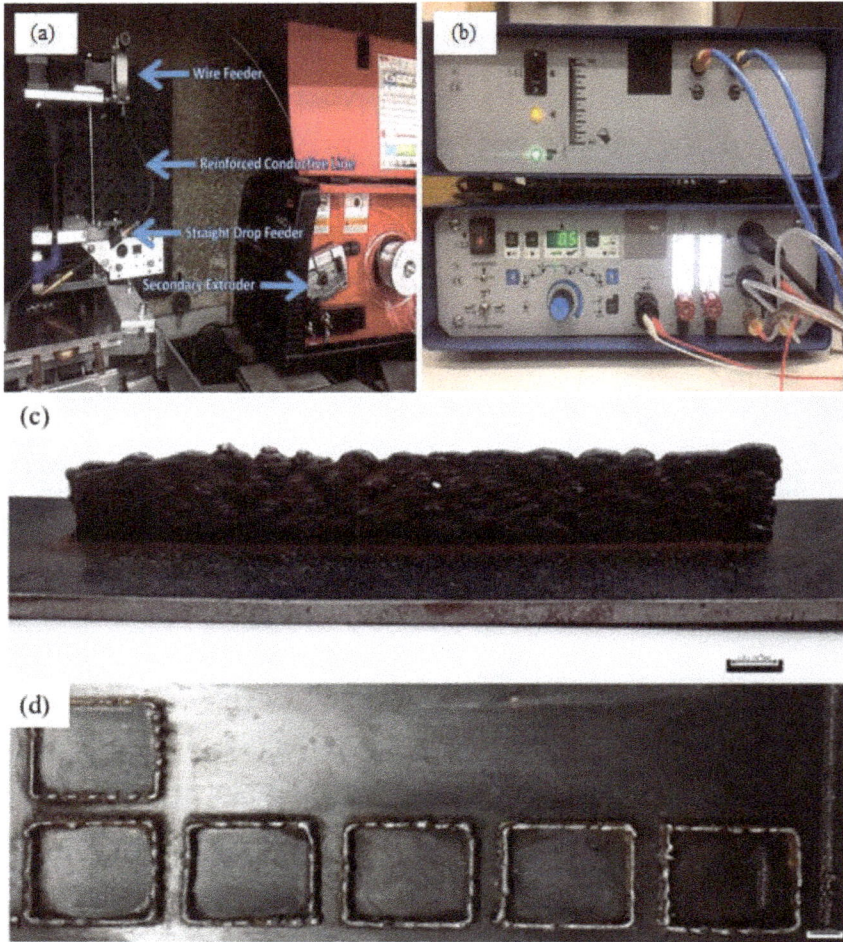

**Figure 1.** Experimental setup: (**a**) Overall system for WAAM; (**b**) pro-Fusion Dual Arc 82HFP welding machine used a power supply; (**c,d**) stainless steel produced by WAAM (scale bar represents 1 cm).

## 2.3. Sample Preparation

A series of trials using different printing parameters were performed to determine ideal printing parameters. Parameters related with the intensity of the arc were first found by a trial-and-error method. It was found that one of the main factors that affect the quality of the line was the gas flow rate of the pilot arc. When the flow rate is low (<0.4 L/min) the arc was not able to melt the wire. On the other hand, high flow rates (>0.8 L/min) create a strong jet that carves the substrate and ejects the molten metal wire away from the molten pool. Thus, a medium pilot arc gas flow rate was used (0.6 L/min). A shielding gas flow rate of 12 L/min was used for all the trials, protecting the melting pool from the exposed air. For both the pilot gas and the shielding gas 99.99% Argon was used. The cross-section area of the selected wire's diameter was 1.5 mm and 12 V motor was used. The minimum feed rate that the motor is capable to produce is 117 cm/min (46 in/min). This feed rate was then used for all the trials.

Samples were prepared at various arc currents and printing speed, keeping all other parameters constant. Table 3 shows a visual representation of the combination of parameters that were used on the trials to determine the best printing conditions. Printing speed was changed from 90 to 180 mm/min, using intervals of 15 mm/min between each trial. 20–50 Ampere were tried. This time, the intervals were kept on 5 A between each trial. A window of parameters in which visually uniform lines could be printed was found. 15 parameters inside this printing window were selected (5 current intensities; 30 A, 32.5 A, 35 A, 37.5 A, and 40 A, and 3 printing speeds 120 mm/min, 135 mm/min, and 150 mm/min). Using each one of these combinations, five parallel lines were printed over the same substrate. After printing, the lines were detached from the substrate and their dimensions were measured. The lines were sectioned and mounted in epoxy resin for further analysis and characterization. All the samples were wet ground through successive grades of silicon carbide abrasive papers from P240 to P4000 (Buehler, Lake Bluff, IL, USA) and etched with glycol (1 mL $HNO_3$, 24 mL of water, 75 mL Ethylene Glycol) [20].

**Table 3.** Combination of parameters during printing.

| Current (Ampere) | Printing Speed (mm/min) | Heat Input (kJ/mm) |
|---|---|---|
| 30 | 120 | 0.225 |
| 30 | 135 | 0.200 |
| 30 | 150 | 0.180 |
| 32.5 | 120 | 0.244 |
| 32.5 | 135 | 0.217 |
| 32.5 | 150 | 0.195 |
| 35 | 120 | 0.263 |
| 35 | 135 | 0.233 |
| 35 | 150 | 0.210 |
| 37.5 | 120 | 0.281 |
| 37.5 | 135 | 0.250 |
| 37.5 | 150 | 0.225 |
| 40 | 120 | 0.300 |
| 40 | 135 | 0.267 |
| 40 | 150 | 0.240 |

*2.4. Experimental Design*

2.4.1. Optical Microscope

Micrographs of the magnesium wire as well as the WAAM produced samples were taken for visual comparison. Additional images were taken of WAAM samples after the corrosion test to identify any changes. The optical microscopes used for this project were Nikon and Eclipse LV15ONL (Nikon, Brighton, MI, USA) and KEYENCE VHX-900 Digital Microscope (KEYENCE, Itasca, IL, USA) equipped with a VH-Z100UR lens.

2.4.2. X-ray Diffraction (XRD)

A Bruker XRD D8 system (BRUKER AXS, Madison, WI, USA) using Cu $K9$ (1 = 1.54 Å) X-ray source was used to for the analysis of the as-received wire and printed lines.

2.4.3. Scanning Electron Microscopy (SEM) and Energy Dispersive Spectroscopy (EDS)

The use of SEM was limited to obtaining reference images for EDS mapping. A Hitachi S4800 (Hitachi, Dallas, TX, USA) equipped with EDS was used. Two different areas of AZ91D produced by WAAM the cross-section area were analyzed (center and top). The accelerating voltage was 10 kV for all the cases.

2.4.4. Electrochemical Impedance Spectroscopy (EIS)

The Electrochemical Impedance Spectroscopy was carried out by using a computer controlled potentiostat (VersaSTAT 3F, AMETEK SI, Oak Ridge, TN, USA) with 3.5 wt % NaCl solution at room temperature. The experimental setup consisted of three electrode cell containing graphite as counter electrode, 3D printed sample as working electrode and saturated Ag/AgCl electrode (saturated with KCl, Accumet Glass Body (Fisher Scientific, Dallas, TX, USA)) was used as the reference electrode. Working electrodes, the electrical contact between a copper wire and the sample was made at the back of the sample with silver conductive epoxy. The whole assembly was protected from the solution by mounting it in epoxy resins, which makes it leave only one face of the electrode exposed to the solution. The exposed surface area was 0.77 cm$^2$.

## 3. Results and Discussion

### 3.1. Sample Preparation

Figure 2 shows representative lines printed over a range of printing conditions. Low temperature conditions can be achieved by using fast moving speed and applying low arc current, such as in the case of condition A. In this case, it is noticeable that the wire did not completely melt. Traces of the original wire can be seen in many points of the line. When dealing with low currents (20–30 A) but low printing speeds, such as the case of condition B, mounds of fully molten metal can be seen. It is believed that the high amount of materials deposited due to the low speeds made created the partition of the original line. On the contrary cases, when high currents (>40 A) the metal wire becomes highly fluid. Thus, at printing speeds greater than 170 mm/min (condition D), the molten metal will flow and congregate on the beginning of the printed line, since it is the first solidified region. At lower speeds (<160 mm/min) the molten metal has more time for its solidification. For this reason, the molten metal will flow and conglomerate in those places where the liquid metal has become solid, as seen on conditions E. However, printing conditions in which mostly visually uniform lines was found which was the combination of medium arc currents (30–40 A) and relatively low printing speeds yielded the best results. C shows representative lines printed under conditions laying in this "printing window".

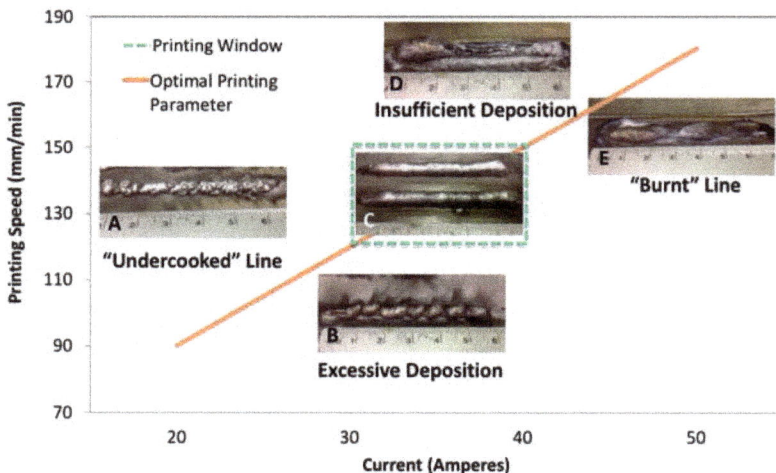

**Figure 2.** Visual representation of the macroscopically appearance of printed lines created under different printing conditions.

Through printing, a decrease in the contact angle as a function of the heat conditions was detected. For all the printing parameters the wire feed rate remained constant. The flow rate of both the pilot arc and the shielding gas was also kept constant. Thus, only two out of the four parameters affecting the contact angle of the printed specimens were varied. First, the size of the weld pool, which is in direct correlation with the current intensity. Second, as the supplied current increases, more heat is delivered to the substrate, increasing the size of the molten pool. The printing speed of the build is based on the movement of the melting pool and not on the wire feed rate. The wire feed rate must be greater to allow enough material to melt and fuse to the plate. Also, by increasing the size of the molten pool (increasing current) the contact angle would decrease since the surface energy differential between the molten substrate and the depositing metal would be lower than that between a solid substrate and a liquid filler. Table 3 shows the heat input into the system depending on the printing conditions. Noticeable is the fact that the amount of heat delivered into the system is directly proportional to the arc current and inversely proportional to the printing speed, to calculate the heat input the following formula was used:

$$Heat\ Input = \frac{Arc\ Voltage * Arc\ Current * Thermal\ Efficiency}{Travel\ speed}$$

where arc voltage is 25 volts (constant for all the cases), thermal efficiency is 60% for PAW, arc current (Ampere) and travel speed (mm/s) are the variables [21].

Due to the relatively low melting point of magnesium and its alloys, remelting is a common problem when printing these materials [22]. Thus, as the temperature of the substrate increases because of the increasing number of lines printed over the same plate, the lines become more susceptible to remelting. It is believed that, along the increased size of the molten pool, remelting is a cause of the reduction of contact angle between the substrate and the printed line as more lines are printed over the same plate. Figure 3 shows two examples of lines presenting remelting. A change in contact angle is noticeable on the areas closest to the substrate. This indicates that heat coming from the printing plate was enough to melt part of the printed line.

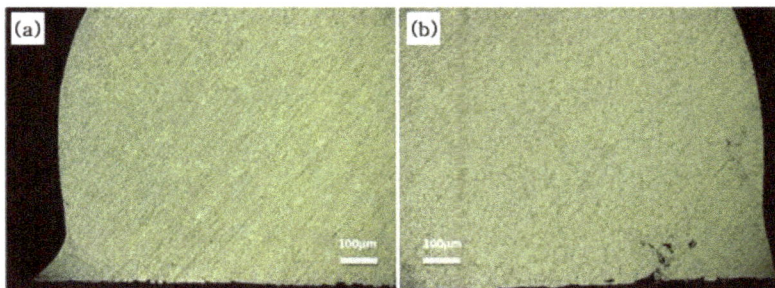

**Figure 3.** Micrographs of two printed lines presenting a decrease in the contact angle due to re-melting: (**a**) Left side of printed line; (**b**) Right side of printed line (Eclipse LV15ONL and KEYENCE VHX-900 Digital Microscope).

### 3.2. Microscopic Analysis

With the use of optical microscopy, the microstructure of the as received AZ91D wire was analyzed. Figure 4 shows a micrograph of a longitudinal cut of the feedstock. A fairly equiaxed grain appears on the micrograph. The presence of equiaxed grains indicates that the wire was rolled at high temperatures and then cooled by air. The slow heat dissipation inside the wire coil made possible conditions like those required for annealing, hence the final microstructure [23]. The presence of second phase particles aligned on the rolling direction can also be seen. The presence of defects was also detected.

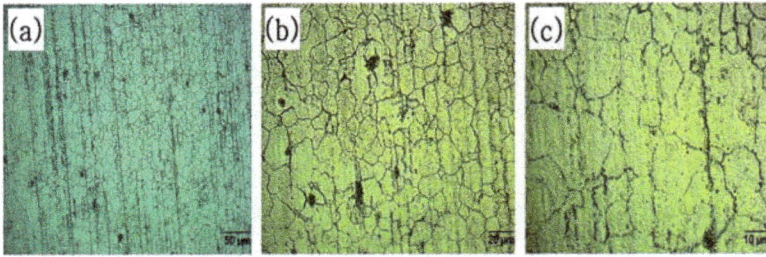

**Figure 4.** Optical micrograph of a longitudinal cot of the as-received AZ91D wire at: (**a**) 20×; (**b**) 50×; (**c**) 100×.

Figure 5 shows a representative microscopy of a printed line, which demonstrates that the microstructure of the printed structure is considerably different than that of the AZ91D wire. High density of well-dispersed second phase particles can be seen in Figure 5a compared to Figure 4a. This microstructure is repeated for all the printed lines regardless the printing parameters or its number within the same printing plate. Both longitudinal and transversal directions show the same arrangement.

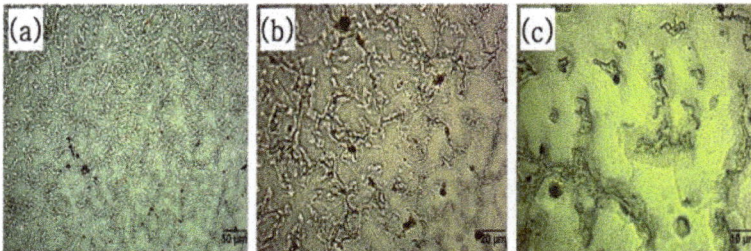

**Figure 5.** Representative microstructure of printed lines. $Al_5Mg_{11}Zn_4$ precipitate (j phase) embedded on a Mg matrix (a phase). (**a**) 20×; (**b**) 50×; (**c**) 100×.

Based on the microscopic analysis, there is an assumption of a precipitate within the magnesium build. The precipitate is assumed to be $Al_5Mg_{11}Zn_4$ based on previous work done within this research department. X-ray diffraction (XRD) confirmed the presence of the precipitate (see Figure 6a,b), while energy dispersive spectroscopy result approved the dispersion of the elements through the material. Figure 6c represents the two different sections where elemental mappings were taken and Figure 6d,e show the result of elemental mappings (see Figure 6d,e).

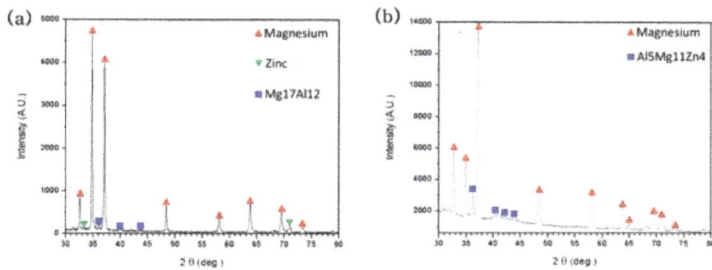

**Figure 6.** *Cont.*

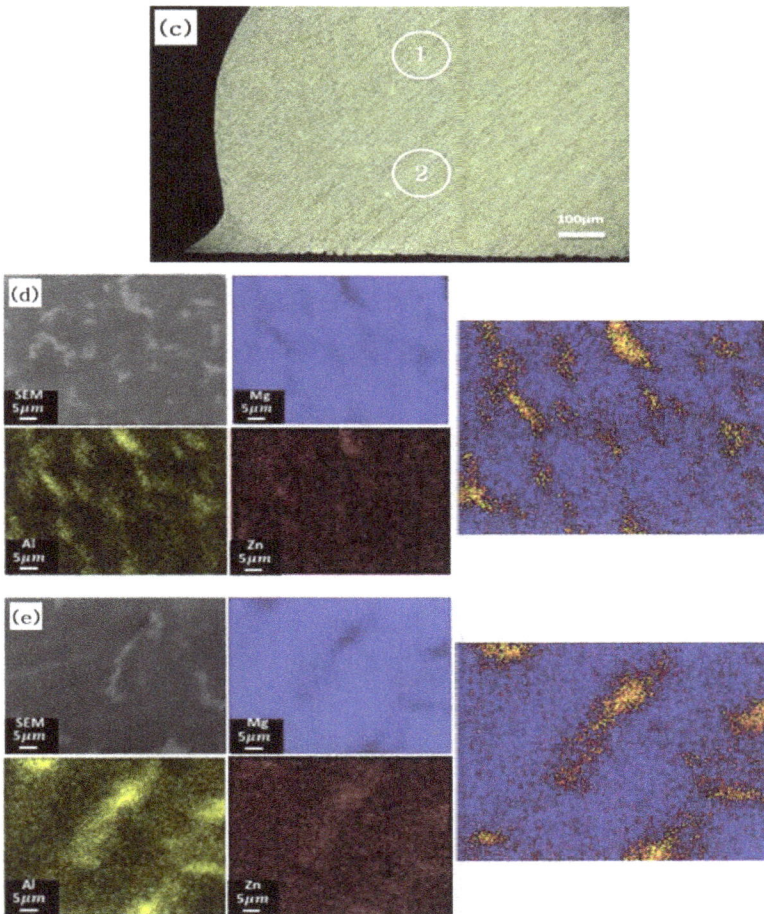

**Figure 6.** (**a**) XRD spectra of as-received wire, representative peaks of magnesium (a phase), zinc, and $Mg_{18}Al_{13}$ precipitates (b phase); (**b**) XRD spectra of a representative printed line, representative peaks of pure magnesium (a phase) and $Al_5Mg_{11}Zn_4$ precipitates (j phase); (**c**) representing the areas from which the elemental mappings were taken; (**d**) Elemental Mapping of zone 1 [24]. Copyright 2018, Cambridge University Press; (**e**) Elemental Mapping of zone 2 [24].

## 3.3. Electrochemical Impedance Spectroscopy (EIS)

EIS graph (see Figure 7) reveals a semi-circular form where its diameter is related to the composition and microstructure. Several authors have described this arc's diameter being associated with corrosion resistance [25,26]. With that said, AZ91D produced by WAAM had six times larger diameter than the AZ91D produced by casting which means corrosion resistance of AZ91D produced by WAAM was better [27]. This result can be related to the new precipitates ($Al_5Mg_{11}Zn_4$) and relieve the internal stress by using WAAM process.

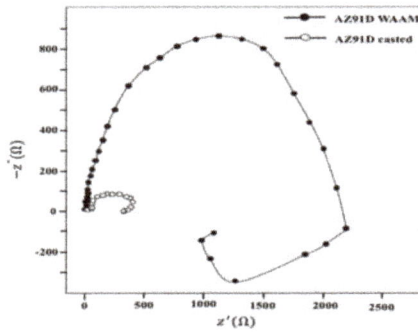

**Figure 7.** EIS results of AZ91D WAAM and AZ91D casted [27] 3.5 wt % NaCl. Copyright 2008, Elsevier.

### 3.4. Microstructure after Corrosion

Figure 8 presents microscopic views of the corroded surface of the as-received wire AZ91D and AZ91D produced by WAAM. Figure 8a produced circular pits which initiated at the magnesium matrix (a phase). On the other hand, Figure 8b shows less pits than as-received wire. This is due to the well-dispersed second phase particles on the surface that act as a corrosion barrier to subsequent barrier.

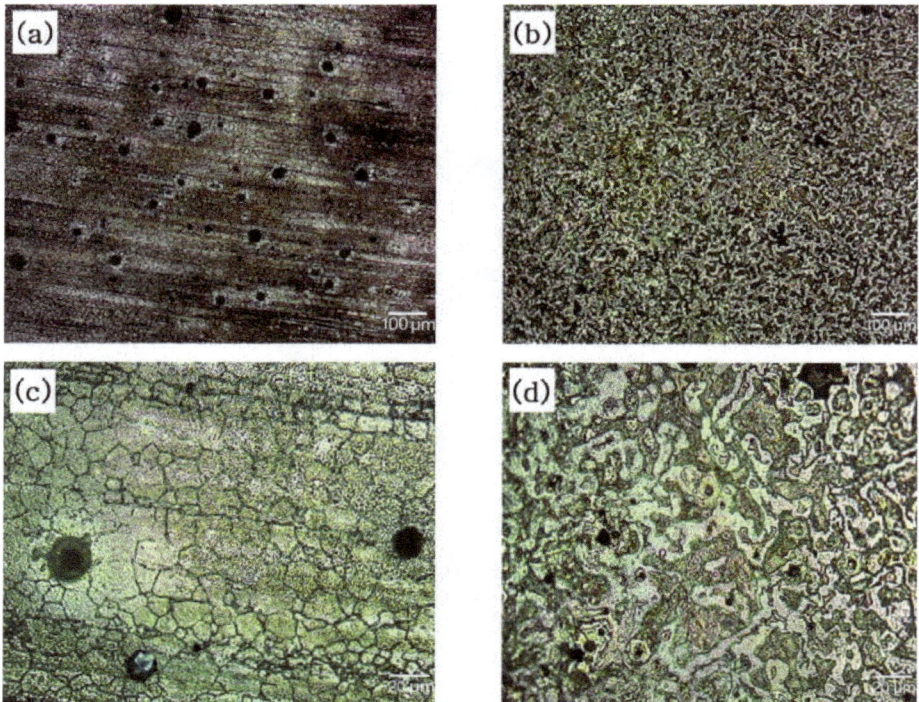

**Figure 8.** After corrosion microstructure of AZ91D produced by WAAM and casting: (**a,c**) AZ91D casted; (**b,d**) AZ91D WAAM.

## 4. Conclusions

In this research, it has been proven that magnesium alloy AZ91D can be used for the creation of specimens using WAAM systems. To obtain uniform base lines, the thermal gradient generated by the plasma arc must be reduced by controlling the movement of the printing bed. Furthermore, the orthorhombic phase present in the microstructure is not what is expected for this alloy. The shielding gas used for removing stagnant air also quenches the sample during the welding process. Zinc atoms are not able to migrate into the bulk crystal, forming $Al_5Mg_{11}Zn_4$ precipitate. This precipitate formation has higher corrosion resistance than a cast magnesium sample.

**Author Contributions:** N.K. conceived and designed the experiments; S.H. designed and performed corrosion experiments; M.Z. contributed corrosion experiment and installed WAAM; D.M.H. designed and installed WAAM; M.M.P. analyzed the microscopic data.

**Funding:** This research was funded by [Korea Institute of Machinery & materials], grant number [CRC-15-03-KIMM].

**Acknowledgments:** This work was supported by (Korea Institute of Machinery & Materialrs; CRC-15-03-KIMM), and this research has emanated from grant SKU-UTEP 2011-0215.

**Conflicts of Interest:** The authors declare no conflict of interest.

## References

1. Rezwan, K.; Chen, Q.Z.; Blaker, J.J.; Boccaccini, A.R. Biodegradable and Bioactive Porous Polymer/Inorganic Composite Scaffolds for Bone Tissue Engineering. *Biomaterials* **2006**, *27*, 3413–3431. [CrossRef] [PubMed]
2. Nájera, S.E.; Michel, M.; Kim, K.-S. 3D Printed PLA/PCL/TiO$_2$ Composite for Bone Replacement and Grafting. *MRS Adv.* **2018**, 1–6. [CrossRef]
3. Niinomi, M.; Hattori, T.; Morikawa, K.; Kasuga, T.; Suzuki, A.; Fukui, H.; Niwa, S. Development of Low Rigidity β-Type Titanium Alloy for Biomedical Applications. *Mater. Trans.* **2002**, *43*, 2970–2977. [CrossRef]
4. Ridzwan, M.I.Z.; Shuib, S.; Hassan, A.Y.; Shokri, A.A.; Ibrahim, M.N.M. Problem of stress shielding and improvement to the hip implant designs: A review. *J. Med. Sci.* **2007**, *7*, 460–467. [CrossRef]
5. Lhotka, C.; Szekeres, T.; Steffan, I.; Zhuber, K.; Zweymüller, K. Four-Year Study of Cobalt and Chromium Blood Levels in Patients Managed with Two Different Metal-on-Metal Total Hip Replacements. *J. Orthop. Res.* **2003**, *21*, 189–195. [CrossRef]
6. Staiger, M.P.; Pietak, A.M.; Huadmai, J.; Dias, G. Magnesium and Its Alloys as Orthopedic Biomaterials: A Review. *Biomaterials* **2006**, *27*, 1728–1734. [CrossRef] [PubMed]
7. Witte, F.; Eliezer, A. Biodegradable Metals. In *Degradation of Implant Materials*; Eliaz, N., Ed.; Springer: New York, NY, USA, 2012; pp. 93–109, ISBN 978-1-4614-3941-7.
8. Li, N.; Zheng, Y. Novel Magnesium Alloys Developed for Biomedical Application: A Review. *J. Mater. Sci. Technol.* **2013**, *29*, 489–502. [CrossRef]
9. Witte, F.; Fischer, J.; Nellesen, J.; Crostack, H.; Kaese, V.; Pisch, A.; Beckmann, F.; Windhagen, H. In vitro and in vivo Corrosion Measurements of Magnesium Alloys. *Biomaterials* **2006**, *27*, 1013–1018. [CrossRef] [PubMed]
10. Levesque, J.; Dube, D.; Fiset, M.; Mantovani, D. Investigation of Corrosion Behaviour of Magnesium Alloy AM60B-F under Pseudo-Physiological Conditions. *Mater. Sci. Forum* **2003**, *424–426*, 521–526. Available online: https://www.cheric.org/research/tech/periodicals/view.php?seq=1254436 (accessed on 16 July 2018). [CrossRef]
11. Xin, Y.; Liu, C.; Zhang, X.; Tang, G.; Tian, X.; Chu, P.K. Corrosion Behavior of Biomedical AZ91 Magnesium Alloy in Simulated Body Fluids. *J. Mater. Res.* **2007**, *22*, 2004–2011. [CrossRef]
12. Gao, M.; Wei, R.P. A 'Hydrogen Partitioning' Model for Hydrogen Assisted Crack Growth. *Metall. Trans. A* **1985**, *16*, 2039–2050. [CrossRef]
13. Regev, M.; Aghion, E.; Rosen, A.; Bamberger, M. Creep Studies of Coarse-Grained AZ91D Magnesium Castings. *Mater. Sci. Eng. A* **1998**, *252*, 6–16. [CrossRef]
14. Wong, K.V.; Hernandez, A. A Review of Additive Manufacturing. *ISRN Mech. Eng.* **2012**, *2012*, 1–10. [CrossRef]
15. Frazier, W.E. Metal Additive Manufacturing: A Review. *J. Mater. Eng. Perform.* **2014**, *23*, 1917–1928. [CrossRef]
16. Han, K.N.; Kim, N.S. Challenges and Opportunities in Direct Write Technology Using Nano-Metal Particles. *KONA Powder Part. J.* **2009**, *27*, 73–83. [CrossRef]

17. Amert, A.K.; Oh, D.-H.; Kim, N.-S. A Simulation and Experimental Study on Packing of Nanoinks to Attain Better Conductivity. *J. Appl. Phys.* **2010**, *108*, 102806. [CrossRef]

18. Kim, N.S.; Han, K.N.; Church, K.H. Direct Writing Technology for 21th Century Industries—Focus on Micro—Dispensing Deposition Write Technology. In Proceedings of the Korean Society of Machine Tool Engineers Spring Conference 2007, Seoul, Korea, 14–19 May 2007.

19. Taminger, K.M.B.; Hafley, R.A. Electron Beam Freeform Fabrication: A Rapid Metal Deposition Process. In Proceedings of the 3rd Annual Automotive Composites Conference, Troy, MI, USA, 9–10 September 2003.

20. George, B.; Voort, V. Metallography of Magnesium and Its Alloys. *Buehler Tech-Notes* **2015**, *4*, 1–5.

21. Huo, H.; Li, Y.; Wang, F. Corrosion of AZ91D Magnesium Alloy with a Chemical Conversion Coating and Electroless Nickel Layer. *Corros. Sci.* **2004**, *46*, 1467–1477. [CrossRef]

22. West, E.G. *The Welding of Non-Ferrous Metals*, 1st ed.; John Wiley & Sons: New York, NY, USA, 1951.

23. Humpherys, F.J. *Recrystallization and Related Annealing Phenomena*, 1st ed.; Elsevier B. V.: Kidlington, Oxford, UK, 1995.

24. Martinez, D.A.; Han, S.; Kim, N.P. Magnesium Alloy 3D Printing by Wire and Arc Additive Manufacturing (WAAM). *MRS Adv.* **2018**, 1–6. [CrossRef]

25. Cao, C.-N. On the Impedance Plane Displays for Irreversible Electrode Reactions Based on the Stability Conditions of the Steady-State—I. One State Variable besides Electrode Potential. *Electrochim. Acta* **1990**, *35*, 831–836. [CrossRef]

26. Cao, C.-N. On the Impedance Plane Displays for Irreversible Electrode Reactions Based on the Stability Conditions of the Steady-State—II. Two State Variables besides Electrode Potential. *Electrochim. Acta* **1990**, *35*, 837–844. [CrossRef]

27. Pardo, A.; Merino, M.C.; Coy, A.E.; Viejo, F.; Arrabal, R.; Feliú, S. Influence of Microstructure and Composition on the Corrosion Behaviour of Mg/Al Alloys in Chloride Media. *Electrochim. Acta* **2008**, *53*, 7890–7902. [CrossRef]

applied
sciences

MDPI

*Article*

# Superior Mechanical Behavior and Fretting Wear Resistance of 3D-Printed Inconel 625 Superalloy

**Yong Gao [1] and Mingzhuo Zhou [2,\*]**

1. Wuhan Institute of Marine Electric Propulsion, Wuhan 430064, China; 18907120357@163.com
2. The State Key Laboratory of Digital Manufacturing Equipment and Technology, School of Mechanical Science and Engineering, Huazhong University of Science and Technology (HUST), Wuhan 430074, China
* Correspondence: mingzhuo@hust.edu.cn; Tel.: +86-132-6067-8851

Received: 17 October 2018; Accepted: 16 November 2018; Published: 1 December 2018

**Abstract:** Additive manufacturing (AM) nickel-based superalloys have been demonstrated to equate or exceed mechanical properties of cast and wrought counterparts but their tribological potentials have not been fully realized. This study investigates fretting wear behaviors of Inconel 625 against the 42 $CrMo_4$ stainless steel under flat-on-flat contacts. Inconel 625 is prepared by additive manufacturing (AM) using the electron beam selective melting. Results show that it has a high hardness (335 HV), superior tensile strength (952 MPa) and yield strength (793 MPa). Tribological tests indicate that the AM-Inconel 625 can suppress wear of the surface within a depth of only ~2.4 μm at a contact load of 106 N after $2 \times 10^4$ cycles. The excellent wear resistance is attributed to the improved strength and the formation of continuous tribo-layers containing a mixture of $Fe_2O_3$, $Fe_3O_4$, $Cr_2O_3$ and $Mn_2O_3$.

**Keywords:** nickel; fretting; hardness; wear mechanisms

## 1. Introduction

Since the first 3D printing process appeared in the Japanese edition of the IEICE Transactions on Electronics in 1981, explorations on different metal components from metallic powders have rapidly increased worldwide for investigating high-performance metallic materials or complex structures in the fundamental scientific research. It is noteworthy that the additive manufacturing (AM) is becoming the focus of intensive research owing to its advantages of completely melting the powders and producing fully dense microstructures during the fabrication without pretreatment, thus making them a superior technique in AM processes. In addition to the ability of AM to create complex structures with a high density, the capability to prepare materials with an ultra-high tensile strength (TS) and a superior yield strength (YS) has also been demonstrated. The excellent TS reported for AM-processed steel, titanium and nickel-based alloy are 1470 [1], 1510 [2] and 1184 MPa [3], respectively, with corresponding YS of 1100, 1440 and 933 MPa.

As one of the most important nickel-based alloys, Inconel 625 has shown exceptional properties such as strong resistance to corrosion, high tensile strength and yield strength, as well as high resistance to chloride-ion stress-corrosion cracking [4,5] and thus has been widely applied in advanced areas including sea-water applications. It has been used as propeller blades for motor patrol gunboats, submarine auxiliary propulsion motors and submarine quick-disconnect fittings [4]. Potential applications are springs, seals, fasteners and oceanographic instrument components. Conventionally, Inconel 625 alloys are hard-to-cut materials during machining, which normally require high cutting temperatures due to the low thermal conductivity and volume-specific heat. Compared with traditional casting and forging parts, Inconel 625 obtained by the AM process shows much higher tensile strength [6,7] and has hardness as high as 343 HV with primary dendrite arm space of ~0.5 μm due to the rapid cooling speed (106 K/s) [6]. These unique properties make AM processes very attractive for designing the high-strength and durable Inconel 625 applied in aerospace, automotive and robotics.

However, Inconel 625 superalloy applied in moving assemblies normally subjects to various critical conditions such as cyclic fatigue loading, corrosion and thermal stresses and thus suffers from serious mechanical energy dissipation caused by friction and wear. Accordingly, a fine control of friction and wear for Inconel 625 superalloy is desirable for improving the reliability of moving mechanical systems and is also an attractive fundamental scientific research.

Iwabuchi [8] first investigated fretting properties of Inconel 625 under different pressures at elevated temperature and high vacuum. The friction coefficient with a high value of 1.7 at pressures below 0.1 Pa was observed and it decreased to a low value of 0.55 at 105 Pa. In order to enhance the wear resistance, the reinforced Inconel 625 with considerably size-reduced TiC particles was successfully prepared by Hong et al. [9] using the laser metal deposition (LMD). During the LMD process, the in-situ interfacial reaction could be tailored between the TiC reinforcement and the Ni-based metal matrix to improve the interfacial wettability and bonding coherence and thus greatly enhancing the wear resistance. Recently, an accurate 3D finite element model was developed by Lotfi et al. [10] to predict the wear of Inconel 625 superalloy, acting as tools with complex geometries, with PVD-TiAlN coated carbide and ceramic inserts. It was found that the low-cutting speed, low-depth cut and the middle-level feed rate would be appropriate for Inconel 625 with PVD-TiAlN coated carbide insert, while ceramic insert demonstrated better performance under the condition of a high-cutting speed.

Recently, increasing attentions on AM metallic materials have been received for designing components with a high complexity. However, data of tribological performance of Inconel 625 superalloy fabricated by AM are very limited. It has been demonstrated that alloys prepared by AM showed higher mechanical properties surpassing their traditional wrought or cast counterparts [11–15]. Herein, Inconel 625 superalloy is prepared in this work by electron beam selective melting (EBSM). Mechanical and tribological behaviors are systemically investigated via an AG-Xplus 100 kN mechanical test system and a flat-on-flat tribo-system, respectively.

## 2. Experiments

### 2.1. Material Preparation

The initial powders selected for this study were commercial Inconel 625 produced by gas atomization with the particle sizes ranging from 22 to 47 μm (as shown in Figure 1) and used without further processing. The data were collected based on the standard method in Reference [16]. The powder is composed of Ni-Cr-Fe-Nb-Mn-Mo-Si. The chemical compositions are given in Table 1. In the present study, Inconel 625 samples were fabricated on a 250 × 250 mm building platform by EBSM R2 device with the beam power of 300 W and scan speeds of 500 mm/s. The processing values were demonstrated as the optimum parameters for manufacturing AM-nickel based super alloys with almost fully density. Figure 2 represents an overview of the AM processing strategy. Each layer was scanned once using vectors oriented along either the $X$-axis (scan direction-x, SD$_x$) or the $Y$-axis (scan direction-y, SD$_y$), alternatively. The fabricated parts were wire-cut from the steel substrate for mechanical polishing with emery papers down to 1200 grit and 0.5 μm wet diamond pastes.

**Table 1.** Chemical compositions of the initial powders.

| Elements Mass Ratio | Chemical Composition (wt. %) | | | | | | |
|---|---|---|---|---|---|---|---|
| | Ni | Cr | Fe | Nb | Mo | Mn | Si |
| | Bal. | 20.5 | 3.4 | 3.6 | 8.9 | 0.4 | 0.5 |

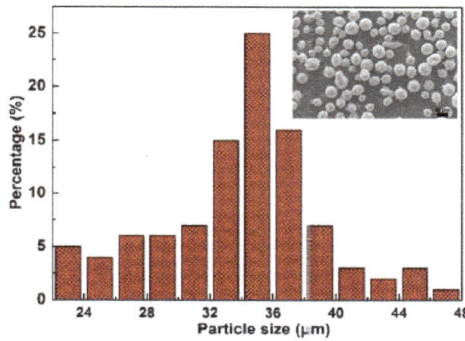

**Figure 1.** Size distribution of initial particles fabricated by the gas atomization.

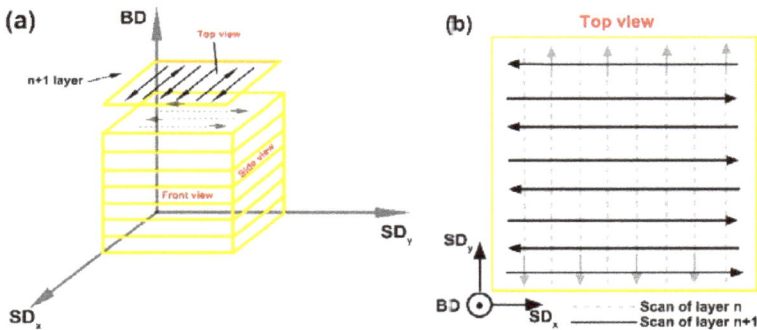

**Figure 2.** Overview of the AM processing strategy. The build (BD) and scan directions (SD$_x$ and SD$_y$) are indicated with respect to the sample coordinates: (**a**) the different views of the sample; (**b**) top view of the sample with the bidirectional scan vectors.

### 2.2. Mechanical and Tribological Tests

The Vicker's hardness of AM-processed specimens was measured using a HVS-1000 Vicker's hardness instrument with a load of 1 kg and a dwell time of 10 s. Five random positions for hardness testing are selected on the surfaces of samples and the mean value was obtained. The density was determined based on the Archimedes principle. Uniaxial tension specimens were extracted from the AM materials by wire electro discharge machining with a gauge dimension of 5.3 mm (length) × 2.9 mm (width) × 1.8 mm (thickness) on this basis of methods in Reference [17]. Samples were clamped by the ends of the plate-type samples. Mechanical properties were assessed by rate-controlled tensile tests at nominal strain rates of $1 \times 10^{-3}$ s$^{-1}$ using a Shimadzu AG-100 kN equipment. All tensile tests were repeated five times.

To evaluate the fretting wear resistance of AM-prepared Inconel 625 alloys, the wear tests were carried out using a fretting test system with a flat-on-flat configuration at contact loads of 40–106 N and a frequency of 2 Hz for 20,000 cycles. Wear test specimens were extracted from the fabricated parts by wire electro discharge machining with a dimension of 30 mm (length) × 30 mm (width) × 7 mm (thickness). Details about wear test configuration can be found in our previous works [18–21]. The experimental parameters were set based on the practical operating conditions of the blade bearing [16]. The contact area of the fretting surface was 6.75 mm$^2$. The angular displacement amplitude ($\theta$) of the fretting test was 1.5°. The 42CrMc4 stainless steel with a hardness of 25 HRC and a roughness of $R_a = 0.2$ μm which is generally used for blade bearing in a propeller was used as a mating material. Tribological experiments were performed at a relative humidity of about 60%. The wear volume was determined by a 3D digital microscope. All the tests were carried out at the same condition for three times to make sure the repeatability of the experimental results at the same conditions and the average results were recorded.

### 2.3. Microstructural Characterization

The crystalline size and phase composition of the initial powders and sintered samples were determined by X-ray diffraction (XRD, Shimadzu Corp., Tokyo, Japan) on a XRD-7000S X-ray diffractometer with Cu K$\alpha$ radiation. A JSM 7600F field-emission scanning electron microscope (FE-SEM, JEOL Ltd., Tokyo, Japan) is used to examine and analyze the microstructures of the wear scars after the friction and wear tests and the fractography after tensile tests with an acceleration voltage of 5 kV and current of 7 µA. The morphologies of the wear scars were also investigated by the 3D optical microscope (OM, Olympus, Tokyo, Japan). The composition of the wear tracks was analyzed by a VG Multilab 2000 X-ray photoelectron spectroscope (XPS, Thermo VG Scientific, Leicestershire, UK).

## 3. Results and Discussion

Metal AM process typically exhibits complex transient material consolidation processes that have an important effect on the resulting grain size and crystal texture [13,14] and thereafter influence mechanical and tribological behaviors of the materials including strength, hardness, friction and wear properties.

### 3.1. Micro-Hardness and Relative Density

Table 2 gives the micro-hardness and relative density of the as-prepared specimen and other Inconel 625 alloys fabricated by different methods such as sintering and welding. Densification result of AM-samples analyzed by the Archimedes principle shows that the relative density of Inconel 625 prepared by AM reaches a value of 99.7%, indicating the printed part from micro-cracks and micro-voids which can promote the increase of the mechanical and tribological performance of materials. Micro-hardness result of AM-samples shows the improved hardness compared to that of the sintering-processed and welding-processed samples. The improvement of the hardness can be ascribed to the residual stresses in AM parts introduced by layer-by-layer fabricating process. For the case of a sufficiently high densification without the formation of micro-cracks or micro-pores, the retention of a reasonable level of residual stress in AM-processed parts favors the enhancement of hardness [15].

**Table 2.** Micro-hardness and relative density of the as-prepared specimen and other Inconel 625 alloys fabricated by different methods.

| Specimen | Method | Micro-Hardness (HV) | Relative Density | Ref. |
|---|---|---|---|---|
| As-prepared | AM | 335 | 99.7% | Present work |
| Inconel 625 | Sintering + ageing | 327 | 99.2% | [16] |
| Inconel 625 | AM | 343 | - | [7] |
| Inconel 625 | Gas tungsten arc deposition | 328 | - | [17] |
| Inconel 625 | Gas tungsten arc welding | 270 | - | [18] |
| Inconel 625 | Sintering | 275 | 91.8 | [19] |
| Milled Inconel 625 | Sintering | 320 | 84 | [19] |
| Inconel 625 | Welding | 252 | - | [20] |

### 3.2. XRD Analyses

The XRD results of the as-received initial Inconel 625 powders and specimens after AM process are shown in the Figure 3. It can be seen that the AM-sample mainly consists of Ni-base austenite phase ($\gamma$ phase) with crystal size of ~5.61 µm. Lattice parameter of the as-received powders is 3.5899 Å and decreases after AM process to 3.5716 Å, which can be demonstrated by a right peak shift on the 2$\theta$

scale for the AM-Inconel 625 (the inset in the Figure 3). Peak shifts are usually caused by the change of the crystal lattice whereas shifts to higher angles are caused by compression of the crystal lattice. This decrease can be attributed to the depletion of Nb, Cr and Mo in the $\gamma$-matrix, since these elements participate in the formation of intermetallic phases [22]. Additionally, a lattice parameter shift in the AM sample compared to the standard alloys 625 fcc $\gamma$-Ni phase also indicates that potentially precipitates in the $\gamma$-Ni matrix are formed during the AM process. However, the phases precipitated in $\gamma$-Ni matrix are not detected by XRD due to a relatively low volume fraction of these precipitates. The result is similar to the previous studies [23]. Only three peaks including $111_\gamma$, $200_\gamma$ and $220_\gamma$ are detected and this reveals the pattern of the columnar dendritic growth [24].

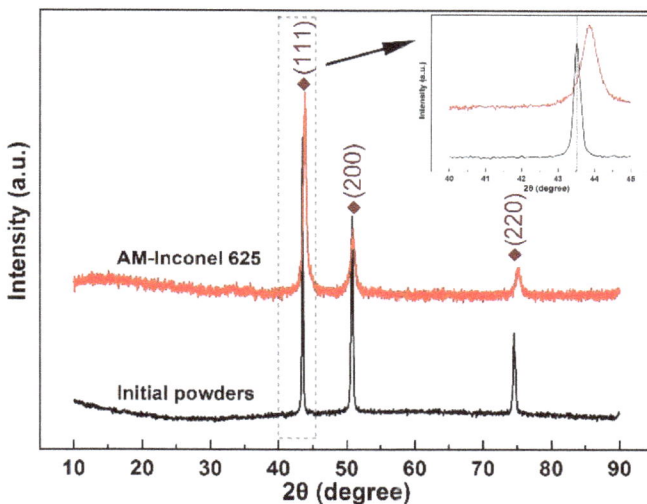

**Figure 3.** XRD results of the as-received initial Inconel 625 powders and specimen after AM process.

*3.3. Tensile Tests*

Figure 4a shows one of the five performed tensile stress-strain curves of Inconel 625 alloy. The inset gives comparison results of the tensile test for Inconel 625 samples in the present work and literatures [25–31]. Apparently, the tensile strength of different processing samples are not identical. Compared with the available data of sintering-processed and welding-processed samples, the ultimate TS of AM-processed Inconel 625 is significantly improved. The sintering and welding samples have a lower ultimate TS (420–830 MPa) compared to that of AM-processed specimens (810–1070 MPa). In this work, Inconel 625 sample prepared by AM have a relatively high ultimate TS of $952 \pm 18$ MPa.

Previous work shows that in general, components made by AM have higher yield and tensile strengths than those of wrought materials of the alloy [32]. The yield strength in equiaxed metals normally relates to the grain size followed by the Hall-Petch low, which can be described as

$$\sigma_y = \sigma_0 + \frac{\kappa}{\sqrt{a}} \tag{1}$$

where $\sigma_y$ is the yield strength, $d$ is the grain size and $\sigma_0$ and $k$ are material constants. To link the yield strength to the grain size of Inconel 625 alloys, yield strength results extracted both from this work and literatures are plotted versus $d^{-0.5}$, as shown in Figure 4b. A linear fit of the yield strength versus relevant grain size gives values of $\sigma_0 = 105$ MPa and $k = 1380$ MPa $\sqrt{\mu m}$ to describe the Hall-Petch relationship between the yield strength and the grain size of Inconel 625. Nevertheless, there are some discrepancies in the predicted yield strength versus $d^{-0.5}$, in which the datum in the present work (793 MPa) does not lie on the line. The value of the yield strength is relative higher than the predicted result, which is likely due to the formation of precipitates in the AM process (Figure 3). Further analysis regarding the precipitate strengthening of the AM-Inconel 625 specimen needs to be carried out after the tensile tests.

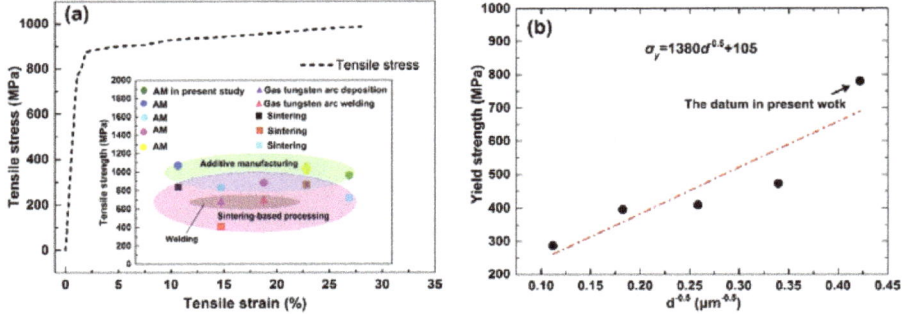

**Figure 4.** (**a**) Tensile stress-strain curve of Inconel 625 alloy and (**b**) yield strengths plotted versus $d^{-0.5}$ for Inconel 625 samples in the present work and literatures, the inset shows the comparison results of tensile strength for Inconel 625 samples in the present work and literatures.

The corresponding fractography of AM-Inconel 625 tested at room temperature is shown in Figure 5. Dimples and particle facets of precipitates are clearly found in the fracture surface, while no obviously microcrack can be observed. This suggests that the failure of AM-Inconel 625 shows a ductile character. A high magnification of the fracture surface (Figure 5b) shows spherical precipitates with a size around 600 nm. In the case of precipitate strengthening ($\Delta\sigma_p$), the dislocation shearing or the Orowan dislocation bypass is believed to be the main strengthening mechanism. For the shearing mechanism, the improvement of the yield strength can be ascribed to contributions of coherency strengthening ($\Delta\sigma_{cs}$), modulus mismatch strengthening ($\Delta\sigma_{ms}$) and order strengthening ($\Delta\sigma_{os}$) [33]. The shearing mechanism is more significant for coherent precipitates with small radii and the transition to Orowan bypass mechanism is active when the radii of coherent precipitates reach a critical value or when the precipitate is separate [34], as can be seen in this work (Figure 5b).

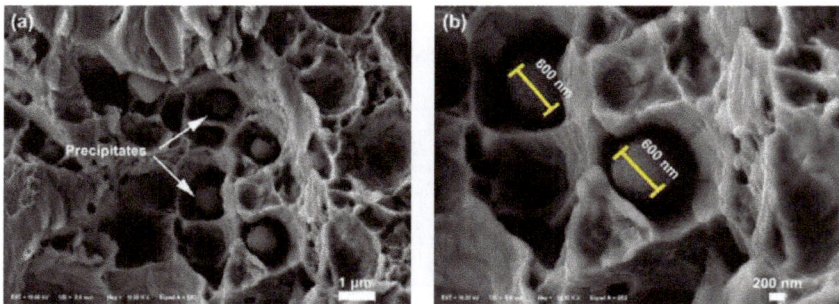

**Figure 5.** (**a**) Fractography of AM-Inconel 625 after tensile test at room temperature; (**b**) High magnification of the fracture surface in (**a**).

For the Orowan dislocation bypass mechanism, the increase in the yield strength can be expressed as

$$\Delta\sigma_{or} = M\frac{0.4Gb}{\pi\lambda}\frac{\ln\left(\frac{2\bar{r}}{b}\right)}{\sqrt{1-\upsilon}} \tag{2}$$

where $M = 3.06$ is the mean orientation factor for the fcc polycrystalline matrix [34]; $G$ is the shear modulus, 81.4 GPa; $v = 0.28$ is Poisson's ratio of the Inconel 625 matrix; $b$ is the magnitude of the Burgers vector, 0.179 nm. For Ni-based superalloys, the dislocations in $\gamma$ phase have a Burgers vector of $a/2$ since precipitates reside within a coherent face-centered cubic (fcc) matrix [35]; $\bar{r} = \sqrt{2/3}r$ is the mean radius of a circular cross section in a random plane for a spherical precipitate, $r$ is the radius of a precipitate, 600 nm. The distribution of spherical precipitates in a cubic grid is applied to simplify a small volume fraction of precipitates, $\lambda$ can be expressed as follows:

$$\lambda = 2\bar{r}\left(\sqrt{\frac{\pi}{4f}} - 1\right) \tag{3}$$

where $f$ is the volume fraction of precipitates, 0.8–1.1%. Thus, the strengthening contribution from the precipitates $\Delta\sigma_p$ is 27.9 MPa.

Since the grain size of the AM-Inconel 625 is significantly reduced, if compared to that of initial powders, the grain boundary (GB) strengthening mechanism is next considered as a possible cause of the YS improvement. It has been established that GB strengthening obeys the Hall-Petch relationship [34]:

$$\Delta\sigma_{GB} = K^{HP}d^{-1/2} \tag{4}$$

where $K^{HP}$ and $d$ are the Hall-Petch coefficient (0.158 MPa m$^{1/2}$, [36] and the average grain diameter (5.61 μm), respectively. The strength increment due to GB, $\Delta\sigma_{GB}$, is 66.7 MPa.

The contributions from different strengthening mechanisms can be added linearly [37,38]. Therefore, the yield strength of the AM-Inconel 625 alloy $\sigma_y$ can be expressed as:

$$\sigma_y = \sigma_0 + \Delta\sigma_{GB} + \Delta\sigma_p \tag{5}$$

where $\sigma_0$ is 687.6 MPa, calculated by the linear relation in Figure 4b. Using the results from Equation, $\sigma_y$ is 782.2 MPa, which is in good agreement with the experimental result of the yield strength (793 MPa). It is worth noting that the full density (100% relative density) is not achieved during the AM process and the presence of a small amount of porosity in the printed samples is not taken into the consideration in strength calculations. Micro-pores may lead to premature failure through the coalescence of micro-pores and the subsequent propagation of cracks, especially for tensile tests.

From the good agreement between the results of the calculation and the experiment, it is reasonable to conclude that the strength of AM-processed Inconel 625 alloys is predictable by quantifying the contributions from different active strengthening mechanisms. The quantification of different strengthening mechanisms can be determined from characterizing the microstructure of Inconel 625 (e.g., distribution and size of spherical precipitates, grain size, etc.), providing a guideline for designing advanced high performance bulk metallic materials by additive manufacturing.

### 3.4. Tribological Tests

Figure 6 shows the relationship between the friction behaviors of AM-Inconel 625 and the number of cycles at a contact load of 106 N. Figure 6a gives friction torques of the material tested after $2 \times 10^4$ cycles. The friction torque increases rapidly over the first $5 \times 10^3$ cycles and reaches a high value (~0.55). Then the friction torque maintains a relative steady state (~0.48). The friction behaviors of AM-Inconel 625 can be more clearly observed in the Figure 6b, in which the friction coefficient of the specimen is plotted against fretting cycles. The friction coefficient reaches a peak value of 0.62 at about $5 \times 10^3$ cycles and decreases to a low value of ~0.50 after $7 \times 10^3$ cycles. The friction coefficient of AM-processed Inconel 625 is similar to that of the traditional Inconel 625 alloy reported in the previous work [8].

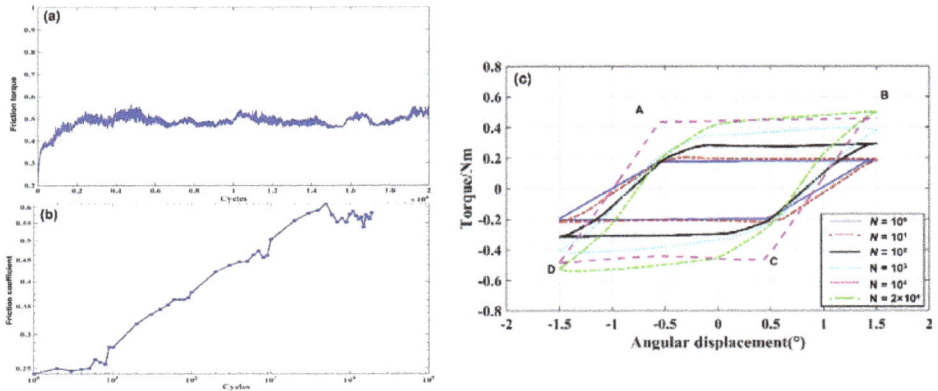

**Figure 6.** Relationship between the (**a**) friction torques and (**b**) friction coefficient of AM-Inconel 625 and the number of cycles at room temperature; (**c**) typical friction torque versus angular displacement amplitude curves (*T-θ* curves) at the fretting cycles of $10^0$, $10^1$, $10^2$, $10^3$, $10^4$ and $2 \times 10^4$.

The typical friction torque versus angular displacement amplitude curves (*T-θ* curves) at the fretting cycles of $10^0$, $10^1$, $10^2$, $10^3$, $10^4$ and $2 \times 10^4$ are displayed in Figure 6c to evaluate the fretting running state. For the *T-θ* curve with a parallelogram shape, the segments $\overline{AB}$ and $\overline{CD}$ correspond to the sliding of the two contact surfaces and the tilt segments $\overline{BC}$ and $\overline{DA}$ correspond to the static friction force. All the *T-θ* curves show the shape of quasi-parallelograms, undoubtedly indicating the gross slip between the contact interfaces during the fretting wear process. For an individual loop, the tangential force is almost constant and without tips at the end, indicating that adhesion wear is mild between the contact surfaces.

To explore the wear mechanism of AM-processed Inconel 625 alloy, optical images of the fretting wear tracks from tests at 40–106 N loads are shown in Figure 7. Figure 7a represents the measurements obtained from the test of AM-processed Inconel 625 at a contact load of 40 N, showing the material transfer from the counterpart. The optical image of the fretting wear track after test at 62 and 84 N are shown in Figure 7b,c, respectively. Similarly, material transfer can be observed on both wear tracks and apparent surface oxidation occurred at a contact load of 84 N. It has been confirmed that fretting wear decreased substantially for most friction couples (metals) as the relative humidity increases from zero to fifty percent [39]. In this study, with humidity about 60%, moisture promotes the formation of soft iron hydrates such as abrasive $Fe_3O_4$, magnetite and a magnetic oxide of iron. Figure 7d shows the wear track obtained after test at a contact load of 106 N. Severe oxidation on the surface could be observed. All the tests at different contact loads last for $2 \times 10^4$ cycles without experiencing a significant amount of surface damage, indicating superior wear-resistance of AM-processed Inconel 625.

Fretting wear is quantified according to a change in wear volume of the fretting wear track, as shown in Figure 8a. All the wear volume tests were carried out at the same condition for three times, yielding standard error bars. The 3D optical microscope is used to measure the wear volume and depth after fretting tests. For testing at 40–86 N, a slight increase in wear volume proves the mild wear in the wear surfaces. Since increment of wear volume is relative pronounced at the highest load of 106 N, the wear still remains slight. The critical load, at which the transition from mild to severe wear occurs, is primarily a material property [40–55]. The propagation of cracks which can result in the significant wear becomes more difficult as the fracture toughness of the material increases. Therefore, it can be inferred that the AM-Inconel 625 would show a high fracture toughness, leading to the superior wear-resistance at varying load condition.

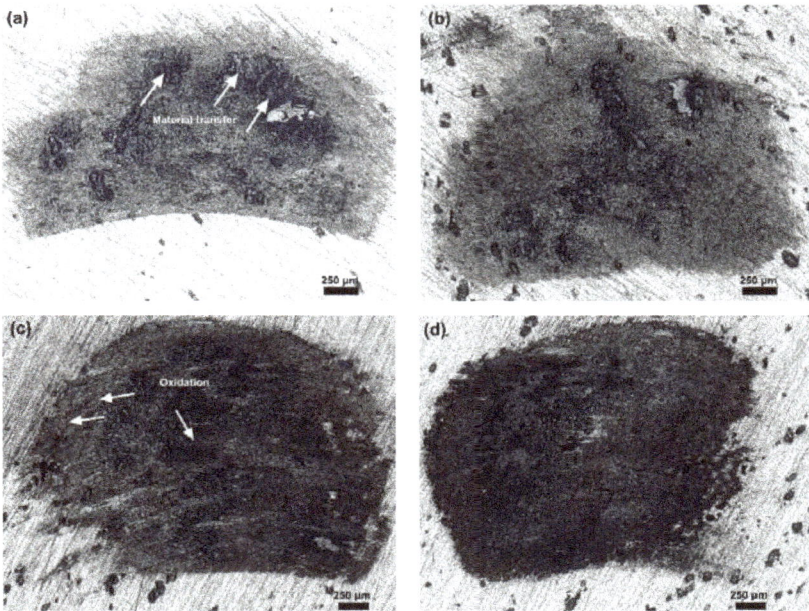

**Figure 7.** Images of the fretting wear tracks of AM-Inconel 625 at (**a**) 40; (**b**) 62; (**c**) 84 and (**d**) 106 N.

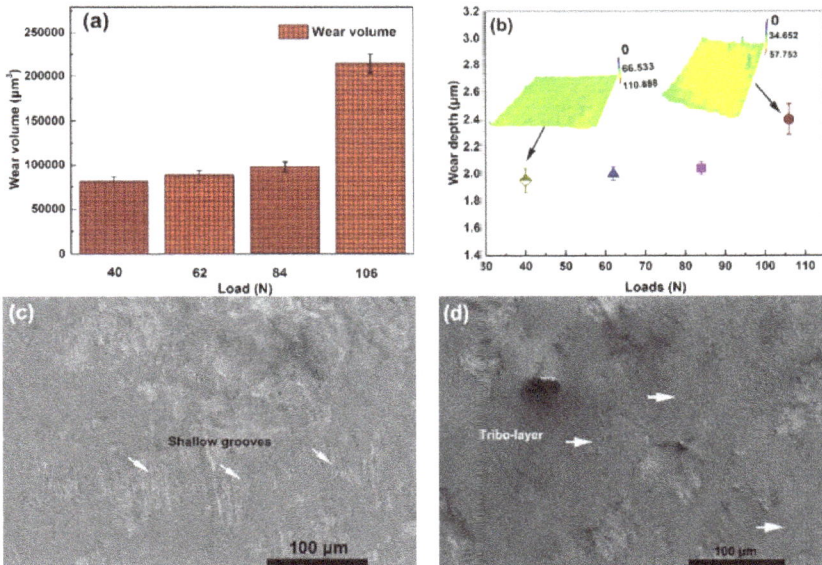

**Figure 8.** (**a**) Wear volume and (**b**) depth of the fretting wear track of AM-Inconel 625 after fretting wear tests; SEM images of the corresponding wear tracks of AM-processed Inconel 625 after tests at (**c**) 40 and (**d**) 106 N; the insets in b show 3D surface topographies of the wear track at 40 and 106 N load, respectively.

During fretting, the wear volume has been proved to be related to the energy dissipated in the contact. Energy dissipation per cycle of fretting can be obtained by:

$$E_f = 2 \int_0^\delta F_s dr = 2F_s\delta = 2\mu F_N \delta \tag{6}$$

where $F_s = \mu \cdot F_N$ is the frictional force, $\mu$ and $F_N$ are the friction coefficient and the normal load, respectively, $\delta$ is the displacement amplitude. With the increased normal load, kinetic energy dissipation increases and the thermal effect plays a significant role in the tribo-oxidation. Meanwhile, the abrasion action of wear debris could be aggravated, leading to the increased wear volume. The calculated values of the dissipation energy with increasing normal load are shown in Table 3.

**Table 3.** Dissipated energy with the increasing normal load.

| Normal Load (N) | Friction Coefficient | Dissipation Energy (J) |
|---|---|---|
| 40 | 0.39 | $47.44 \times 10^{-4}$ |
| 62 | 0.42 | $62.50 \times 10^{-4}$ |
| 84 | 0.46 | $92.74 \times 10^{-4}$ |
| 106 | 0.50 | $127.20 \times 10^{-4}$ |

It can be observed that at the same displacement amplitude, an increase of the normal load leads to the increase of dissipated energy because the dissipated energy is related with both the friction coefficient and the normal load (Equation (6)). This dissipated energy can promote the thermo-mechanical induced transformation of surface layer (such as the transformed layer, plastic flow, delamination) and tribo-oxidation. It was found that the wear volume increased linearly with increasing dissipated energy.

Figure 8b presents the wear depth of wear tracks at different loads, where the quantified wear can be more clearly observed. Each test of the wear depth was carried out at the same condition for three times. An increasing trend can be seen and the maximum depth (~2.4 μm) is obtained at the load of 106 N after $2 \times 10^4$ cycles. The wear depth of the commonly used nickel-based protective thin coatings, such as NiTi film [56] usually varies as 40–140 μm per 200 cycles depending on the operating conditions. Thus, it can be demonstrated that the AM-processed Inconel 625 possess extraordinary anti-wear property, which suppresses wear of the fretting contact surface for as long as $2 \times 10^4$ cycles. The 3D surface topographies of the wear track at 40 and 106 N load are further taken as illustrated in the insets of Figure 8b. No obvious wear can be observed with the AM-processed Inconel 625 alloy.

Detailed SEM images of the corresponding wear tracks of AM-processed Inconel 625 after tests at 40 and 106 N are shown in Figure 8c,d, respectively and the underlying wear mechanisms for the formation of various topographical features are proposed to be as follows. The presence of shallow discontinuous grooves and a small number of abrasive fragments attached on the wear track at 40 N (see Figure 8c) reveals the mild abrasive wear and materials transfer during fretting wear. At a high applied load of 106 N, the wear track of Inconel 625 becomes rather smooth and is found to be covered with an adhesion tribo-layer. In the previous works, microstructural changes and tribochemical reaction are often observed near wear tracks in microcrystalline materials, most commonly in the form of the nanocrystalline tribo-layer [57–66]. Such microstructural refinement can result in a further hardened surface and thus leading to the enhancement of the wear-resistance.

In order to get reliable chemical and structural information from the wear tracks of AM-processed Inconel 625, high-resolution XPS spectra obtained on the tested roller specimens are shown in Figure 9. All the spectra were first calibrated from the C 1s peak at binding energy of 284.4 eV from the NISTXPS database. Additional smoothening was done by the Savitzky-Golay method. The background of the spectra was then subtracted followed by fitting on multiple peaks using the Voigt function. Considering the material transfer from the counterpart (42CrMo4 stainless steel), the Fe 2p core level spectrum of the tribo-layer is shown in Figure 9a, which could be deconvoluted into five components. The peak at 712.2 eV [67] corresponds to $Fe_3O_4$ which is formed through oxidation during the fretting wear, while the peak at 713.3 eV corresponded to $2p_{3/2}$ of $Fe^{5+}$ oxidation state [68]. The Fe $2p_{3/2}$ peak at 717.3 eV suggests the existence of $Fe^{2+}$ and $Fe^{3+}$ in wear tracks [69]. The peak at 724.0 eV [66] is identified as $Fe_2O_3$, while 732.5 eV is $Fe_3O_4$ [21]. The existence of both $Fe_2O_3$ and $Fe_3O_4$ in the tribo-layer confirms that the observed material transfer with tribo-oxidation reaction plays an important role in the formation of the tribo-layer. Figure 9b shows the Cr $2p$ core level spectra for the wear track of AM-processed Inconel 625. The peak at 576.1 eV [70] is identified as $Cr_2O_3$, while the one at 585.7 eV is identified as $Cr^{3+}$ $2p_{1/2}$. The Mn $2p$ spectra of the wear track are shown in Figure 9c. A peak position is obtained at 641.7 eV [71] which corresponds to the oxidation state of $Mn_2O_3$. The aforementioned results demonstrate that the well adhering tribo-layer is mainly composed of mixed oxides. Such layer is normally characterized by a nano-scale structure. Kato and co-workers [72] reported the formation of similar metal oxide nano-particles on the wear track and found that a mild wear regime could be established for those oxides with the higher diffusion coefficients and observed coherent tribo-films. Therefore, the high wear-resistance observed on the AM-processed Inconel 625 shows strong dependence on the good mechanical properties including superior tensile and yield strengths, high hardness, as well as the nature of the tribo-layer formed during fretting wear.

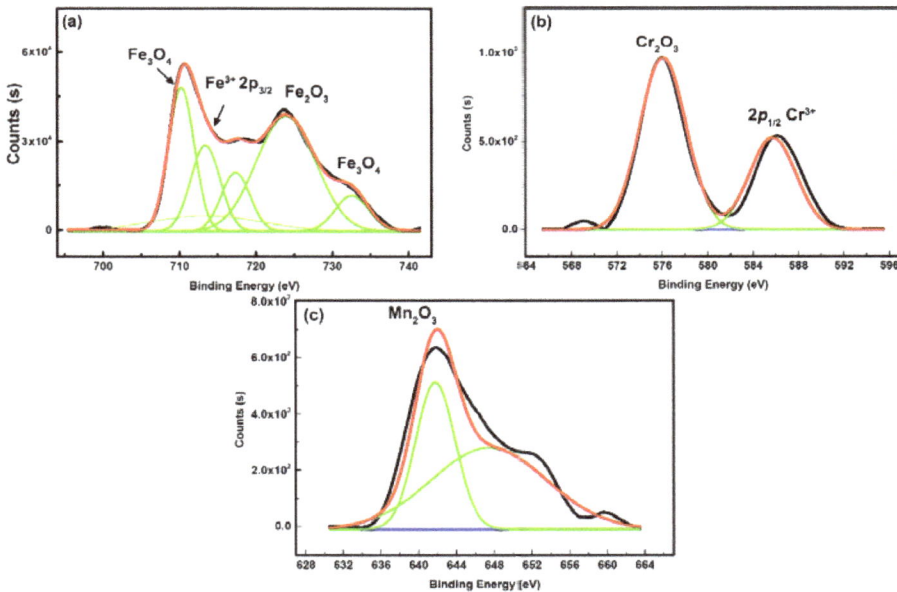

**Figure 9.** High-resolution XPS spectra obtained on the wear tracks of AM-Inconel 625 after fretting wear.

## 4. Conclusions

This study investigates the micro-hardness, tensile strength, yield strength, friction and wear behaviors of AM-processed Inconel 625 alloy to evaluate the effect of AM processing on the mechanical and tribological performance of nickel-based superalloys. AM-processed Inconel 625 shows the significant improvement in tensile and yield strengths, as well as the wear-resistance compared to traditional nickel-based superalloys under the current test conditions. The results concluded from the experimental investigation are as follows:

- Mechanical test results reveal that the AM-processed Inconel 625 have high hardness value (335 HV), superior tensile strength (952 ± 18 MPa) and yield strength (793 MPa), as well as excellent wear-resistance.

*Appl. Sci.* **2018**, *8*, 2439

- Contributions of precipitates and GB strengthening can be quantitatively calculated for the AM-Inconel 625 alloy. The precipitates and GB contribute 27.9 and 66.7 MPa, respectively, to the overall strength.
- The wear depth of traditional nickel-based materials usually varies 40–140 μm per 200 cycles depending on the operating conditions. However, the AM-processed Inconel 625 alloy demonstrates the good wear-resistance, which controls wear depth of the fretting surface within 2.4 μm after $2 \times 10^4$ cycles.
- The wear characteristics of AM-processed Inconel 625 depend significantly on the continuous tribo-layers. XPS analysis shows that the observed stable and coherent tribo-layer is characterized by a mixture of $Fe_2O_3$, $Fe_3O_4$, $Cr_2O_3$ and $Mn_2O_3$.

**Author Contributions:** Y.G. and M.Z. developed the methodology. Y.G. leads the project. M.Z. prepared samples. Y.G. and M.Z. carried out the experiments. Y.G. and M.Z. analyzed the data. M.Z. wrote the manuscript. Y.G. and M.Z. revised the manuscript.

**Funding:** This research was funded by [the National Key R&D Program of China] grant number [2017YFB0103000].

**Conflicts of Interest:** The authors declare no conflict of interest.

## References

1. Spierings, A.B.; Starr, T.L.; Wegener, K. Fatigue performance of additive manufactured metallic parts. *Rapid Prototyp. J.* **2013**, *19*, 88–94. [CrossRef]
2. Chlebus, E.; Kuźnicka, B.; Kurzynowski, T.; Dybała, B. Microstructure and mechanical behaviour of Ti–6Al–7Nb alloy produced by selective laser melting. *Mater. Charact.* **2011**, *62*, 488–495. [CrossRef]
3. Wang, F. Mechanical property study on rapid additive layer manufacture Hastelloy® X alloy by selective laser melting technology. *Int. J. Adv. Manuf. Tech.* **2012**, *58*, 545–551. [CrossRef]
4. Paul, C.P.; Ganesh, P.; Mishra, S.K.; Bhargava, P.; Negi, J.; Nath, A.K. Investigating laser rapid manufacturing for Inconel-625 components. *Opt. Laser Technol.* **2007**, *39*, 800–805. [CrossRef]
5. Shankar, V.; Rao, K.B.S.; Mannan, S.L. Microstructure and mechanical properties of Inconel 625 superalloy. *J. Nucl. Mater.* **2001**, *288*, 222–232. [CrossRef]
6. Yadroitsev, I.; Thivillon, L.; Bertrand, P.; Smurov, I. Strategy of manufacturing components with designed internal structure by selective laser melting of metallic powder. *Appl. Surf. Sci.* **2007**, *254*, 980–983. [CrossRef]
7. Li, S.; Wei, Q.S.; Shi, Y.S.; Zhu, Z.C.; Zhang, D.Q. Microstructure Characteristics of Inconel 625 Superalloy Manufactured by Selective Laser Melting. *J. Mater. Sci. Technol.* **2015**, *31*, 946–952. [CrossRef]
8. Iwabuchi, A. Fretting wear of inconel 625 at high temperature and in high vacuum. *Wear* **1985**, *106*, 163–175. [CrossRef]
9. Hong, C.; Gu, D.; Dai, D.; Alkhayat, M.; Urban, W.; Yuan, P.; Cao, S.; Gasser, A.; Weisheit, A.; Kelbassa, I. Laser additive manufacturing of ultrafine TiC particle reinforced Inconel 625 based composite parts: Tailored microstructures and enhanced performance. *Mater. Sci. Eng. A* **2015**, *635*, 118–128. [CrossRef]
10. Lotfi, M.; Jahanbakhsh, M.; Farid, A.A. Wear estimation of ceramic and coated carbide tools in turning of Inconel 625: 3D FE analysis. *Tribol. Int.* **2016**, *99*, 107–116. [CrossRef]
11. Sames, W.; List, F.; Pannala, S.; Dehoff, R.; Babu, S. The metallurgy and processing science of metal additive manufacturing. *Int. Mater. Rev.* **2016**, *61*, 1–46. [CrossRef]
12. Lewandowski, J.J.; Seifi, M. Metal Additive Manufacturing: A Review of Mechanical Properties. *Annu. Rev. Mater. Res.* **2016**, *46*, 151–186. [CrossRef]
13. Basak, A.; Das, S. Epitaxy and Microstructure Evolution in Metal Additive Manufacturing. *Annu. Rev. Mater. Res.* **2016**, *46*, 527–530. [CrossRef]
14. Ramsperger, M.; Mújica Roncery, L.; Lopez-Galilea, I.; Singer, R.F.; Theisen, W.; Körner, C. Solution Heat Treatment of the Single Crystal Nickel-Base Superalloy CMSX-4 Fabricated by Selective Electron Beam Melting. *Adv. Eng. Mater.* **2015**, *17*, 1486–1493. [CrossRef]
15. Gu, D.; Meiners, W. Microstructure characteristics and formation mechanisms of in situ WC cemented carbide based hardmetals prepared by Selective Laser Melting. *Mater. Sci. Eng. A* **2010**, *527*, 7585–7592. [CrossRef]

16. Calleja, A.; Tabernero, I.; Fernández, A.; Celaya, A.; Lamikiz, A.; López de Lacalle, L.N. Improvement of strategies and parameters for multi-axis laser cladding operations. *Opt. Laser Eng.* **2014**, *56*, 113–120. [CrossRef]

17. Martínez Krahmer, D.; Polvorosa, R.; López de Lacalle, L.N.; Alonso-Pinillos, U.; Abatel, G.; Riu, F. Alternatives for Specimen Manufacturing in Tensile Testing of Steel Plates. *Exp. Tech.* **2016**, *40*, 1555–1565. [CrossRef]

18. Zhai, W.; Lu, W.; Liu, X.; Zhou, L. Nanodiamond as an effective additive in oil to dramatically reduce friction and wear for fretting steel/copper interfaces. *Tribol. Int.* **2019**, *129*, 75–81. [CrossRef]

19. Zhai, W.; Lu, W.; Zhang, P.; Wang, J.; Liu, X.; Zhou, L. Wear-triggered self-healing behavior on the surface of nanocrystalline nickel aluminum bronze/$Ti_3SiC_2$ composites. *Appl. Surf. Sci.* **2018**, *436*, 1038–1049. [CrossRef]

20. Zhai, W.; Lu, W.; Chen, Y.; Liu, X.; Zhou, L.; Lin, D. Gas-atomized copper-based particles encapsulated in graphene oxide for high wear-resistant composites. *Compos. Part B* **2019**, *157*, 131–139. [CrossRef]

21. Zhou, M.; Lu, W.; Liu, X.; Zhai, W.; Zhang, P.; Zhang, G. Fretting wear properties of plasma-sprayed $Ti_3SiC_2$ coatings with oxidative crack-healing feature. *Tribol. Int.* **2018**, *118*, 196–207. [CrossRef]

22. Mostafaei, A.; Toman, J.; Stevens, E.L.; Hughes, E.T.; Krimer, Y.L.; Chmielus, M. Microstructural evolution and mechanical properties of differently heat-treated binder jet printed samples from gas- and water-atomized alloy 625 powders. *Acta Mater.* **2016**, *122*, 280–289. [CrossRef]

23. Tian, Y.; Gontcharov, A.; Gauvin, R.; Lowden, P.; Brochu, M. Effect of heat treatment on microstructure evolution and mechanical properties of Inconel 625 with 0.4 wt % boron modification fabricated by gas tungsten arc deposition. *Mater. Sci. Eng. A* **2016**, *684*, 275–283. [CrossRef]

24. Wang, J.F.; Sun, Q.J.; Wang, H.; Liu, J.P.; Feng, J.C. Effect of location on microstructure and mechanical properties of additive layer manufactured Inconel 625 using gas tungsten arc welding. *Mater. Sci. Eng. A* **2016**, *676*, 395–405. [CrossRef]

25. Wang, P.; Li, T.; Lim, Y.F.; Tan, C.K.I.; Chi, D. Sintering and mechanical properties of mechanically milled Inconel 625 superalloy and its composite reinforced by carbon nanotube. *Met. Powder Rep.* **2016**. [CrossRef]

26. Li, G.; Huang, J.; Wu, Y. An investigation on microstructure and properties of dissimilar welded Inconel 625 and SUS 304 using high-power $CO_2$ laser. *Int. J. Adv. Manuf. Technol.* **2015**, *76*, 1203–1214. [CrossRef]

27. Jalal, M.; Ramezanianpour, A.A.; Pool, M.K. Split tensile strength of binary blended self compacting concrete containing low volume fly ash and $TiO_2$ nanoparticles. *Compos. Part B* **2013**, *55*, 324–337. [CrossRef]

28. Gupta, R.K.; Anil Kumar, V.; Gururaja, U.V.; Shivaram, B.R.N.V.; Maruti Prasad, Y.; Ramkumar, P.; Chakravarthi, K.V.A.; Sarkar, P. Processing and Characterization of Inconel 625 Nickel Base Superalloy. *Mater. Sci. Forum* **2015**, *38*, 830–831. [CrossRef]

29. Ma, D.; Stoica, A.D.; Wang, Z.; Beese, A.M. Crystallographic texture in an additively manufactured nickel-base superalloy. *Mater. Sci. Eng. A* **2017**, *684*, 47–53. [CrossRef]

30. Ashtiani, H.R.R.; Zarandooz, R. Microstructural and mechanical properties of resistance spot weld of Inconel 625 supper alloy. *Int. J. Adv. Manuf. Technol.* **2016**, *84*, 607–619. [CrossRef]

31. Hehr, A.; Dapino, M.J. Interfacial shear strength estimates of NiTi–Al matrix composites fabricated via ultrasonic additive manufacturing. *Compos. Part B* **2015**, *77*, 199–208. [CrossRef]

32. Wang, Z.; Palmer, T.A.; Beese, A.M. Effect of processing parameters on microstructure and tensile properties of austenitic stainless steel 304L made by directed energy deposition additive manufacturing. *Acta Mater.* **2016**, *110*, 226–235. [CrossRef]

33. Lloyd, D.J. Precipitation Hardening. *Metall. Mater. Trans. A* **1985**, *16*, 2131–2165.

34. Courtney, T.H. *Mechanical Behavior of Materials*; Waveland Press: Long Grove, IL, USA, 2005.

35. Crudden, D.J.; Mottura, A.; Warnken, N.; Raeisinia, B.; Reed, R.C. Modelling of the influence of alloy composition on flow stress in high-strength nickel-based superalloys. *Acta Mater.* **2014**, *75*, 356–370. [CrossRef]

36. Thompson, A.A. Yielding in nickel as a function of grain or cell size. *Acta Metall.* **1975**, *23*, 1337–1342. [CrossRef]

37. Chen, X.H.; Lu, L.; Lu, K. Grain size dependence of tensile properties in ultrafine-grained Cu with nanoscale twins. *Scr. Mater.* **2011**, *64*, 311–314. [CrossRef]

38. Ma, K.; Wen, H.; Hu, T.; Topping, T.D.; Isheim, D.; Seidman, D.N.; Lavernia, E.J.; Schoenung, J.M. Mechanical behavior and strengthening mechanisms in ultrafine grain precipitation-strengthened aluminum alloy. *Acta Mater.* **2014**, *62*, 141–155. [CrossRef]

39. Basu, B.; Vitchev, R.G.; Vleugels, J.; Celis, J.P.; Biest, O.V.D. Influence of humidity on the fretting wear of self-mated tetragonal zirconia ceramics. *Acta Mater.* **2000**, *48*, 2461–2471. [CrossRef]

40. Wang, J.; Luo, X.H.; Sun, Y.H. Torsional Fretting Wear Properties of Thermal Oxidation-Treated Ti$_3$SiC$_2$ Coatings. *Coatings* **2018**, *8*, 324. [CrossRef]

41. Tewari, A.; Basu, B.; Bordia, R.K. Model for fretting wear of brittle ceramics. *Acta Mater.* **2009**, *57*, 2080–2087. [CrossRef]

42. Swaminathan, V.; Gilbert, J.L. Fretting corrosion of CoCrMo and Ti$_6$Al$_4$V interfaces. *Biomaterials* **2012**, *33*, 5487–5503. [CrossRef] [PubMed]

43. Sikdar, K.; Shekhar, S.; Balani, K. Fretting wear of Mg–Li–Al based alloys. *Wear* **2014**, *318*, 177–187. [CrossRef]

44. Lan, P.; Meyer, J.L.; Vaezian, B.; Polycarpou, A.A. Advanced polymeric coatings for tilting pad bearings with application in the oil and gas industry. *Wear* **2016**, *354*, 10–20. [CrossRef]

45. Zhang, P.; Liu, X.; Lu, W.; Zhai, W.; Zhou, M.; Wang, J. Fretting wear behavior of CuNiAl against 42CrMo4 under different lubrication conditions. *Tribol. Int.* **2018**, *117*, 59–67. [CrossRef]

46. Zhang, P.; Lu, W.; Liu, X.; Zhai, W.; Zhou, M.; Jiang, X. A comparative study on torsional fretting and torsional sliding wear of CuNiAl under different lubricated conditions. *Tribol. Int.* **2018**, *117*, 78–86. [CrossRef]

47. Zhang, P.; Lu, W.; Liu, X.; Zhai, W.; Zhou, M.; Zeng, W. Torsional fretting and torsional sliding wear behaviors of CuNiAl against 42CrMo$_4$ under dry condition. *Tribol. Int.* **2018**, *118*, 11–19. [CrossRef]

48. Lu, W.; Zhai, W.; Zhang, P.; Zhou, M.; Liu, X.; Zhou, L. Effect of different levels of free water in oil on the fretting wear of nickel-aluminum bronze based composites. *Wear* **2017**, *390–391*, 376–384. [CrossRef]

49. Zhai, W.; Lu, W.; Zhang, P.; Zhou, M.; Liu, X.; Zhou, L. Microstructure, mechanical and tribological properties of nickel-aluminium bronze alloys developed via gas-atomization and spark plasma sintering. *Mater. Sci. Eng. A* **2017**, *707*, 325–336. [CrossRef]

50. Zhang, P.; Lu, W.; Liu, X.; Zhou, M.; Zhai, W.; Zhang, G.; Zeng, W.; Jiang, X. Torsional fretting wear behavior of CuNiAl against 42CrMo4 under flat on flat contact. *Wear* **2017**, *380*, 6–14. [CrossRef]

51. Zhang, P.; Lu, W.; Liu, X.; Zhou, M.; Zhai, W.; Zhang, G.; Zeng, W.; Jiang, X.; Lu, W.; Zhang, P.; et al. Influence of surface topography on torsional fretting wear under flat-on-flat contact. *Tribol. Int.* **2017**, *109*, 367–372.

52. Zhai, W.; Shi, X.; Yang, K.; Huang, Y.; Zhou, L.; Lu, W. Mechanical and tribological behaviors of the tribo-layer of the nanocrystalline sturcture during sliding contact: Experiments and model assessment. *Compos. Part B* **2017**, *108*, 354–363. [CrossRef]

53. Zhai, W.; Shi, X.; Yang, K.; Zhou, L.; Lu, W. Tribological Behaviors of Ni$_3$Al Intermetallics with MoO$_3$ Multilayer Ribbon Crystal Prepared by Spark Plasma Sintering. *Acta Metall. Sin Engl.* **2017**, *30*, 576–584. [CrossRef]

54. Lu, W.; Zhang, G.; Liu, X.; Jiang, X. Prediction of surface topography at the end of sliding running-in wear based on areal surface parameters. *Tribol. Trans.* **2014**, *57*, 553–560. [CrossRef]

55. Yang, K.; Ma, H.; Li, X.; He, Q. The Analysis in In Situ Preparation, Mechanics and Tribology of TiAl-SnAgCu/Graphene Composites. *Adv. Eng. Mater.* **2018**, *800719*, 1–8. [CrossRef]

56. Ng, K.L.; Sun, Q.P.; Tomozawa, M.; Miyazak, S. Wear behavior of NITI thin film at micro-scale. *Int. J. Mod. Phys. B* **2012**, *24*, 85–93. [CrossRef]

57. Shi, W.; Luo, X.; Zhang, Z.; Liu, Y.; Lu, W. Influence of external load on the frictional characteristics of rotary model using a molecular dynamics approach. *Comput. Mater. Sci.* **2016**, *122*, 201–209. [CrossRef]

58. Lan, P.; Zhang, Y.; Dai, W.; Polycarpou, A. A phenomenological elevated temperature friction model for viscoelastic polymer coatings based on nanoindentation. *Tribol. Int.* **2018**, *119*, 299–307. [CrossRef]

59. Lan, P.; Polychronopoulou, K.; Zhang, Y.; Polycarpou, A.A. Three-body abrasive wear by (silica) sand of advanced polymeric coatings for tilting pad bearings. *Wear* **2017**, *382*, 40–50. [CrossRef]

60. Lan, P.; Gheisari, R.; Meyer, J.; Polycarpou, A. Tribological performance of aromatic thermosetting polyester (ATSP) coatings under cryogenic conditions. *Wear* **2018**, *398*, 47–55. [CrossRef]

61. Lan, P.; Meyer, J.L.; Economy, J.; Polycarpou, A.A. Unlubricated tribological performance of aromatic thermosetting polyester (ATSP) coatings under different temperature conditions. *Tribol. Lett.* **2016**, *61*, 10.

62. Zhang, G.; Liu, X.; Lu, W. A parameter prediction model of running-in based on surface topography. *Proc. Inst. Mech. Eng. Part C* **2013**, *227*, 1047–1055. [CrossRef]

63. Zhang, H.; Brown, L.; Blunt, L.; Jiang, X.; Barrans, S. The contribution of the micropores in bone cement surface to generation of femoral stem wear in total hip replacement. *Tribol. Int.* **2011**, *44*, 1476–1482. [CrossRef]

64. Zhang, H.; Brown, L.T.; Blunt, L.A.; Jiang, X.; Barrans, S.M. Understanding initiation and propagation of fretting wear on the femoral stem in total hip replacement. *Wear* **2009**, *266*, 566–569. [CrossRef]

65. Lan, P.; Polycarpou, A. High temperature and high pressure tribological experiments of advanced polymeric coatings in the presence of drilling mud for oil & gas applications. *Tribol. Int.* **2018**, *120*, 218–225.

66. Shakoor, R.A.; Waware, U.S.; Ali, K.; Kahraman, R.; Popelka, A.; Yusuf, M.M.; Hasan, A. Novel Electrodeposited Ni-B/Y2O3 Composite Coatings with Improved Properties. *Coatings* **2017**, *7*, 161. [CrossRef]

67. Tan, B.J.; Klabunde, K.J.; Sherwood, P.M.A. X-ray photoelectron spectroscopy studies of solvated metal atom dispersed catalysts. Monometallic iron and bimetallic iron-cobalt particles on alumina. *Chem. Mater.* **2002**, *2*, 186–191. [CrossRef]

68. Brion, D. Etude par spectroscopie de photoelectrons de la dégradation superficielle de $FeS_2$, $CuFeS_2$, ZnS et PbS a l'air et dans l'eau. *Appl. Surf. Sci.* **1980**, *5*, 133–152. [CrossRef]

69. Kar, P.K.; Singh, G. Evaluation of Nitrilotrimethylene Phosphonic Acid and Nitrilotriacetic Acid as Corrosion Inhibitors of Mild Steel in Sea Water. *ISRN Mater. Sci.* **2011**, *2011*, 167487. [CrossRef]

70. Grohmann, I.; Kemnitz, E.; Lippitz, A.; Unger, W.E.S. Curve fitting of Cr 2p photoelectron spectra of $Cr_2O_3$ and $CrF_3$. *Surf. Int. Anal.* **2004**, *23*, 887–891. [CrossRef]

71. Allen, G.C.; Harris, S.J.; Jutson, J.A.; Dyke, J.M. A study of a number of mixed transition metal oxide spinels using X-ray photoelectron spectroscopy. *Appl. Surf. Sci.* **1989**, *37*, 111–134. [CrossRef]

72. Kato, H.; Komai, K. Tribofilm formation and mild wear by tribo-sintering of nanometer-sized oxide particles on rubbing steel surfaces. *Wear* **2007**, *262*, 36–41. [CrossRef]

*applied*
*sciences*

MDPI

Article

# Degradation Classification of 3D Printing Thermoplastics Using Fourier Transform Infrared Spectroscopy and Artificial Neural Networks

**Sung-Uk Zhang** [ORCID]

Electro Ceramic Center, Department of Automotive Engineering, Dong-Eui University, 176 Eomgwangro, Busanjin-gu, Busan 47340, Korea; zsunguk@deu.ac.kr

Received: 23 June 2018; Accepted: 21 July 2018; Published: 25 July 2018

**Abstract:** Fused deposition modeling (FDM) is the most popular technology among 3D printing technologies because of inexpensive and flexible extrusion systems with thermoplastic materials. However, thermal degradation phenomena of the 3D-printed thermoplastics is an inevitable problem for long-term reliability. In the current study, thermal degradation of 3D-printed thermoplastics of ABS and PLA was studied. A classification methodology using deep learning strategy was developed so that thermal degradation of the thermoplastics could be classified using FTIR and Artificial Neural Networks (ANNs). Under given data and predefined rules for ANNs, ANN models with nine hidden layers showed the best results in terms of accuracy. To extend this methodology, other thermoplastics, several new datasets for ANNs, and control parameters of ANNs could be further investigated.

**Keywords:** additive manufacturing; ABS; PLA; thermal degradation; FTIR; ANNs

## 1. Introduction

3D printing, also known as additive manufacturing (AM), has been used in automotives, aerospace, mechanical systems, medicine, biological systems, food supply chains, and so on [1,2]. The key advantage of 3D printing is the capability to economically build complex shapes using a wide variety of materials. By using this technology, consumers and industries can rapidly make a prototype in early-stage product design. Seven 3D printing processes have been categorized by ASTM International: material extrusion, powder bed fusion, vat photo-polymerization, material jetting, binder jetting, sheet lamination, and directed energy deposition. Among them, fused deposition modeling (FDM) or fused filament fabrication (FFF), which belongs to the material extrusion process, is becoming the most popular due to its inexpensive and flexible extrusion systems including thermoplastic materials. However, thermal degradation phenomena of the 3D-printed thermoplastics is an inevitable problem for long-term reliability. This study focuses on thermal degradation of the thermoplastics, which is one of the causes of failure.

To characterize thermoplastics and polymers with thermal degradation, researchers have used Fourier transform infrared spectroscopy (FTIR), differential scanning calorimetry (DSC), thermogravimetric analysis (TGA), thermomechanical analysis (TMA), dynamic mechanical analysis (DMA), scanning electron microscopy (SEM), X-ray photoelectron spectroscopy (XPS), and so on. Pan et al. [3] developed a novel system of poly(lactide acid)-D-alpha-tocopheryl polyethylene glycol 1000 succinate (PLA-TPGS) nanoparticles(NPs) for quantum dot (QD) formulation to improve imaging effects. They used FTIR and XPS to analyze the surface chemistry of the samples. Song et al. [4] reported biodegradable lactic-acid-based telechelic pre-polymers incorporating CNTs characterized by FTIR, TGA, DSC, POM, and so on. Wu and Liao [5] reported the production of new biodegradable nanocomposites from polyactide (PLA), tetraethoxysilane (TEOS), and wood flour (WF) using an in-sihi sol-gel process and a melt blending method. They used FTIR to characterize $SiO_2$, PLA, and hybrids.

They found that the PLA-g-AA/SiO$_2$ hybrid could enhance the thermal and mechanical properties of PLA. Chaunier et al. [6] tried to make 3D-printed structure using maize protein as biopolymers. They used DSC and DMA to obtain glass transitions and the thermos-mechanical properties of Zein-based materials and applied direct beam X-ray scattering (WAXS) and FTIR to characterize the structure of Zein during processing. Semba et al. [7] investigated mechanical properties of PLA resin blended with PCL resin. Samples were fabricated using injection molding. They found that the carbonyl groups of the blend material with DCP (dicumyl peroxide) enhanced the viscous property in the PCL phase and the interfacial adhesion in the dual phase nature of the PLA/PCL blend using FTIR spectroscopy. Holland and Hay [8] studied that the thermal degradation of poly(vinyl alcohol) using thermal analysis–FTIR (TA-FTIR), thermogravimetry (TG) and DSC. Liu et al. [9] reported pyrolysis profiles of plastic blends such as ABS/PVC, ABS, PA5, and ABS/PC using thermogravimetric-FTIR (TG-FTIR). Among several methods, FTIR is a nondestructive technique used to obtain an infrared spectrum of absorption or transmission of the thermoplastics. The molecular bond structure could be observed so that material identification, contamination, degradation, and chemical contact are quantitatively characterized using FTIR. However, the information from FTIR is hard to be understood for novices and a feature extraction of FTIR is not easy without a priori knowledge. To overcome these problems, researchers have adopted machine learning or deep learning techniques with FTIR. Ellis et al. [10] used FTIR and the machine learning method to detect microbial spoilage of beef. Argyri et al. [11] used FTIR and artificial neural networks (ANNs) to detect meat spoilage. Sattlecker et al. [12] used FTIR and a support vector machine (SVM) strategy to classify different types and stages of breast cancer.

In this study, thermal degradation of 3D-printed thermoplastics was investigated and classified using FTIR and ANNs. Among the several 3D printing thermoplastics, acrylonitrile butadiene styrene (ABS) and polylactic acid (PLA), which are popular materials for FDM, were used.

## 2. Methodology

### 2.1. Fused Deposition Modeling

Test samples were fabricated by FDM. As mentioned above, two thermoplastics of ABS and PLA were used. The ABS samples were printed using Mojo manufactured by Stratasys. The PLA samples were fabricated using DP201 of Shindoh. The shape of the sample was based on ASTM D638 type IV, which was designed by 3D CAD software. Figure 1 shows the geometry of the sample.

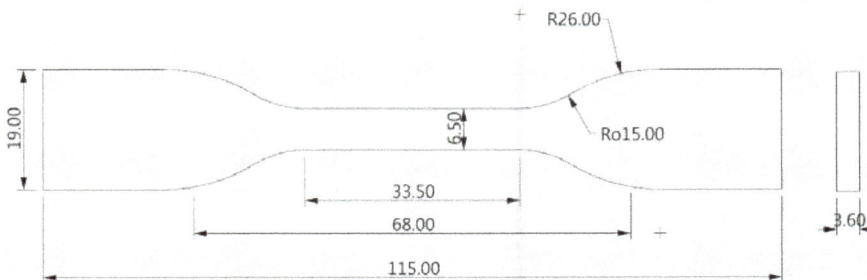

**Figure 1.** ASTM D638 Type IV.

The CAD model was input data for the 3D printers. Infill density was set to be 100%. The printed ABS and PLA specimens are shown in Figure 2. A measuring position for the FTIR was defined to reduce unexpected noises occurred by surface roughness of different measuring positions.

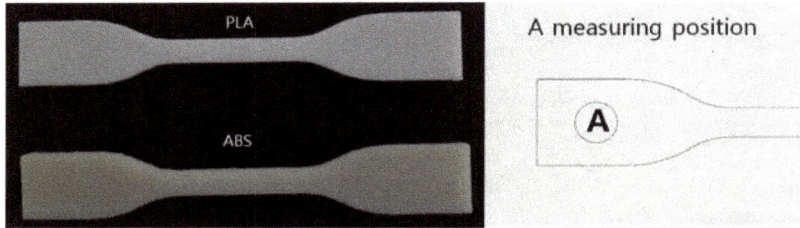

**Figure 2.** Polylactic acid (PLA) and acrylonitrile butadiene styrene (ABS) specimens fabricated by 3D printers and the measuring position for the Fourier transform infrared (FTIR) method.

### 2.2. High-Temperature Storage Test for Polymer Degradation

One time-consuming task is to observe the thermal degradation of thermoplastics at room temperature. To reduce the time, accelerated life testing has been suggested. One accelerated life test is the high-temperature storage test, which was used in this study. The degradation of polymers including thermoplastics is highly sensitive to ambient temperature [13]. SH-662 made of ESPEC was used as a storage chamber. The test can be performed to determine the effects of time and temperature for the degradation phenomenon. In this study, the time and temperature were set to 24 h and 160 °C. Therefore, four groups were prepared: ABS, degraded ABS, PLA, and degraded PLA specimens.

### 2.3. Fourier Transform Infrared Spectroscopy

In this study, a Nicolet iS10 manufactured by Thermo Fisher Scientific (168 Third Avenue Waltham, MA, USA 02451) was used to obtain the infrared spectrum of the 3D-printed ABS and PLA samples. On the predefined position, the FTIR spectrum could be obtained. The number of data points for each FTIR measurement was 6948. The number of repetitions per specimen was 50. Figure 3 shows 50 repetitions per location for an ABS sample. The *x*-axis and *y*-axis represent wavenumber and absorbance.

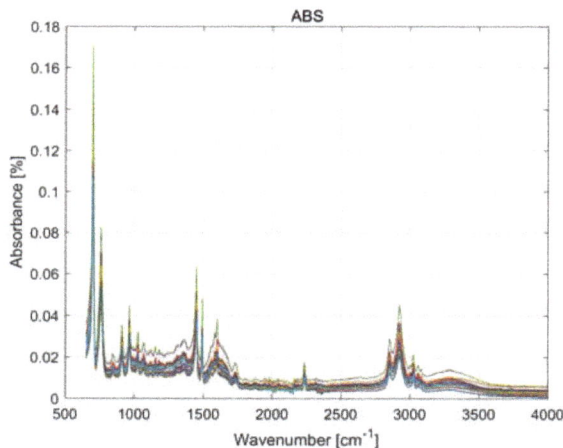

**Figure 3.** The 50 FTIR datasets per a 3D-printed ABS specimen.

Three specimens were prepared for each split. The total number of datasets was 600 for ANNs, as shown in Figure 4. Among all datasets, 10 datasets per split were used as validation sets for the ANNs, and the other datasets were used as training sets.

**Figure 4.** Description of FTIR datasets.

### 2.4. Artificial Neural Networks (ANNs)

ANNs, also called multi-layer perceptions, are computing systems mimicked by biological neural networks, which is graphically shown in Figure 5.

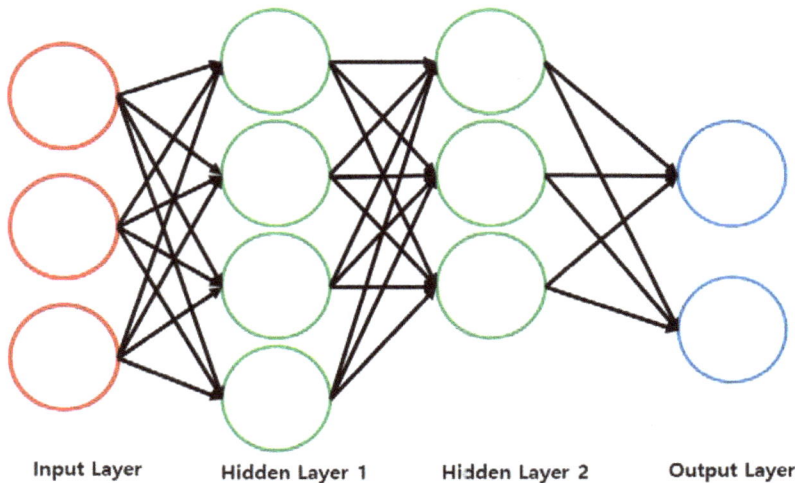

**Figure 5.** A model of artificial neural networks (ANNs).

An ANN is an interconnected group of nodes. The node is called an artificial neuron that receives the input signal, processes it, and transmits the output signal to its neighboring neurons. Signal processing, which is also called as propagation, transformation, a transfer function, or an activation function, is mathematically implemented. Figure 6 shows a sigmoid function (Sigmoid), a hyperbolic function (TanH), and a rectifier linear unit function (ReLU) as the activation function. In this study, the ReLU [14] was used.

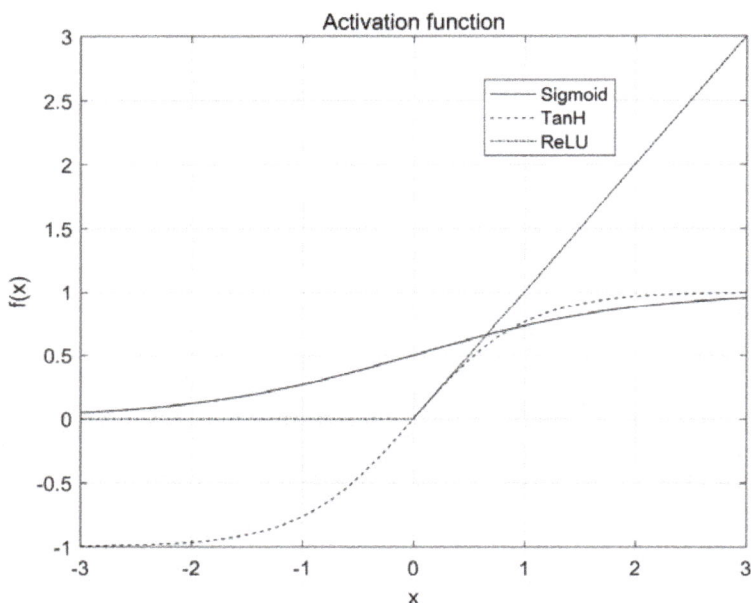

**Figure 6.** Activation functions.

ANNs have three representing layers: the input layer, the hidden layer, and the output layer. The number of hidden layers can be changed. When the number of hidden layers is more than two, the ANNs are defined as a deep neural network. In this study, the numbers of neurons and hidden layers were control parameters during the deep learning process, and this allowed us to observe the tendency toward accuracy for the classification. In addition, the ANN models were generated by the following rules.

- The size of the input layer and that of the output layer are 6948 and 4.
- The size of the hidden layers are between the size of the input layer and that of the output layer.
- The number of hidden neurons is one half of the number of neurons in the previous hidden layer.

## 3. Results and Discussion

### 3.1. Input Datasets for ANNs

After performing the high-temperature storage test, degraded specimens were obtained as shown in Figure 7. The color of the specimens was changed by the accelerated life test. The FTIR measurement was performed on the prescribed position. Figure 8 shows five FTIR measurements per split. In this figure, dABS and dPLA denote the degraded ABS specimen and the degrade PLA specimen. Other researchers have reported characteristic bands for thermal effects. In the case of ABS, the thermal degradation makes the absorption peaks for alkenes and aromatic compounds increase [9]. The characteristic bands of alkenes and aromatic compounds are 3074 cm$^{-1}$, 1630 cm$^{-1}$, 910 cm$^{-1}$, 3033 cm$^{-1}$, 1496 cm$^{-1}$, and 698 cm$^{-1}$. Carrasco et al. [15] reported that the characteristic bands for PLA are related to crystalline structure, which are 1207 cm$^{-1}$, and 920 cm$^{-1}$. However, the IR spectra for PLA and dPLA are hard to be distinguished from each other using classical interpretation methods. When the IR spectra are measured, noise factors involved in the spectra are inevitable, which are typically caused by measurers, the environment, and so on. Noise factors can lead to misinterpretation of spectra. The methodology using ANNs might help to correctly interpret the IR spectra involving various noise factors.

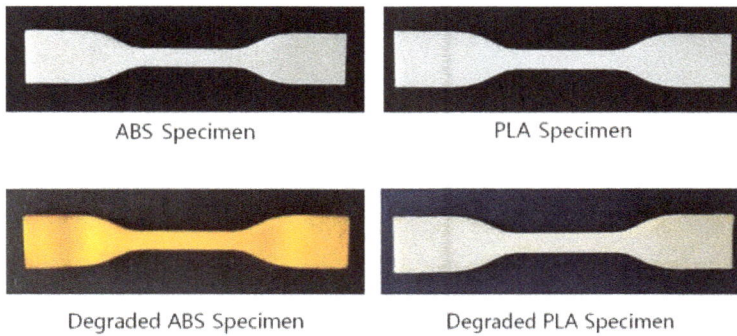

**Figure 7.** Activation functions.

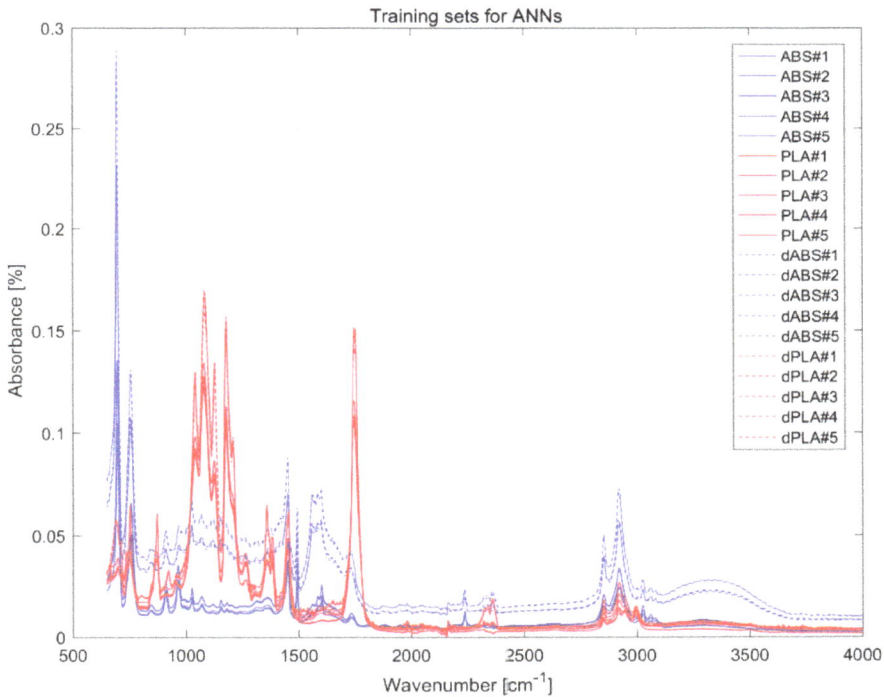

**Figure 8.** Twenty training sets for ANNs.

## 3.2. Validation of ANN Models

Several ANN models were trained using the 560 training sets and validated for 40 test sets. First, ANN models with two hidden layers were validated, as shown in the first graph of Figure 9. In the graph, nH means the number of the hidden layers. The $x$-axis represents the size of the first hidden layer, which is denoted as $H_1$. The $y$-axis describes the average points for the test sets, which is related to the accuracy of the ANN model. When the size of the first hidden layer is 6000, the accuracy of the ANN model is 100%. However, the accuracy of the ANN model becomes 97.5% when the size of the hidden layer is 4000. Similarly, the number of the hidden layers increased from 2 to 11, so all generated ANN models were validated. In the case of the ANN models with 11 hidden layers, the size of the 11th hidden layer became 5 or 6, even though the number of the first layer's hidden neurons was 5000 or 6000. When the number of hidden layers was 9, the accuracies of all generated ANN models were 100%.

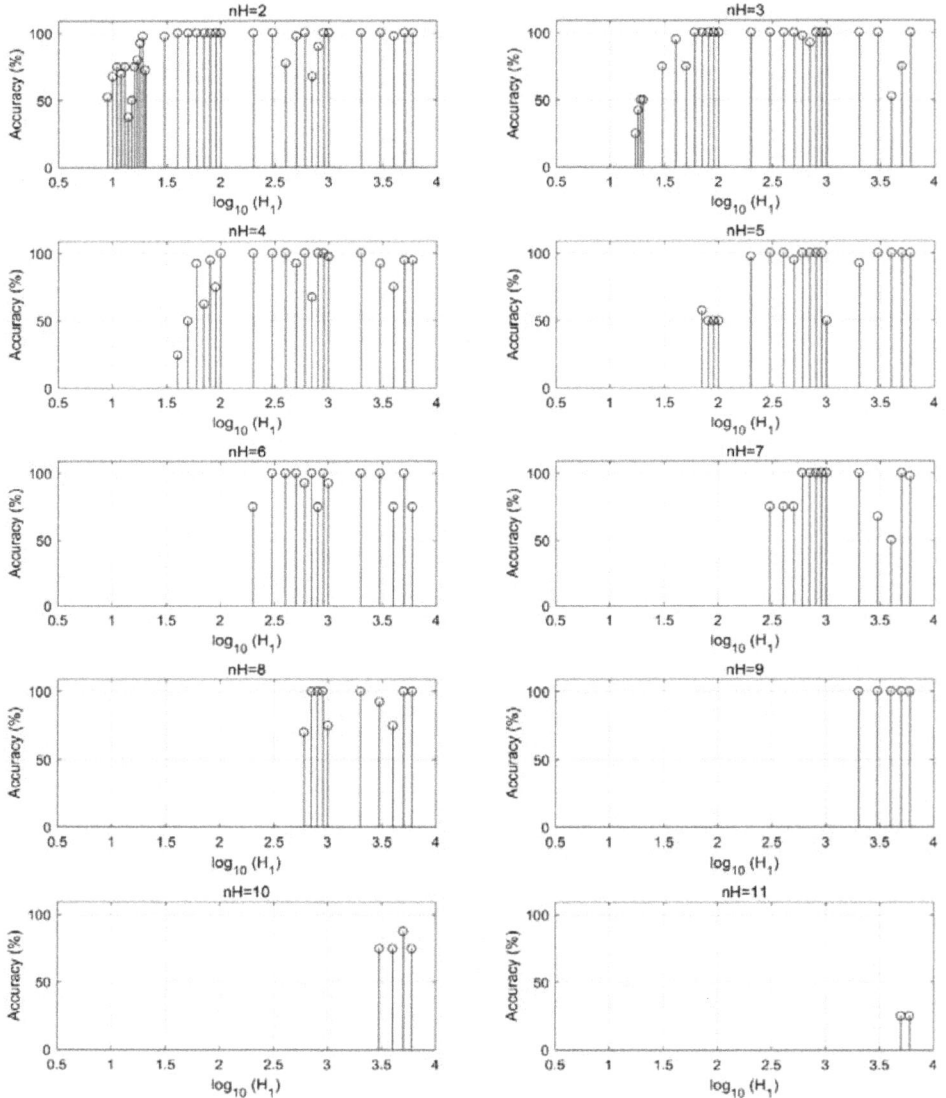

**Figure 9.** Accuracies of ANN models in the change of the number of hidden layers.

*Appl. Sci.* **2018**, *8*, 1224

## 4. Conclusions and Future Work

Under the given data, this study demonstrates that the thermal degradation of the 3D printing thermoplastics of ABS and PLA can be classified using FTIR and ANNs. As the numbers of neurons and hidden layers of ANN models varied, the accuracies of the ANN models were changed. When the ANN models had nine hidden layers, the best results were expected under the given data and rules. To expand and enhance this approach, the other 3D printing thermoplastics of PA, HIPS, TPE, and so on could be applied, test conditions of thermal degradation could be altered in terms of time and temperature, and the numerical parameters of the ANN models could be varied to optimize this methodology. In this study, the high-temperature storage test was performed with only one test condition. The time and the temperature were set to 24 h and 160 °C. To evaluate this methodology more precisely, accelerated life testing under the given conditions should be performed. Two control parameters for ANNs were of focus here, which were the numbers of neurons and hidden layers. As the ANNs are based on optimization, several parameters of this problem, such as learning rate, optimization algorithm, and so on, could act as control factors. Moreover, several activation functions need to be evaluated to optimize this methodology. These issues will be considered in further studies.

**Funding:** This research received no external funding.

**Acknowledgments:** This work was supported by the National Research Foundation of Korea (NRF) grant funded by the Korea government (MSIT) (No. NRF-2017R1C1B5074219).

**Conflicts of Interest:** There are no financial and commercial conflicts of interest in this study.

## References

1. Gao, W.; Zhang, Y.; Ramanujan, D.; Ramani, K.; Chen, Y.; Williams, C.B.; Wang, C.C.L.; Shin, Y.C.; Zhang, S.; Zavattieri, P.D. The status, challenges, and future of additive manufacturing in engineering. *Comput. Des.* **2015**, *69*, 65–89. [CrossRef]
2. Chen, L.; He, Y.; Yang, Y.; Niu, S.; Ren, H. The research status and development trend of additive manufacturing technology. *Int. J. Adv. Manuf. Technol.* **2017**, *89*, 3651–3660. [CrossRef]
3. Pan, J.; Wang, Y.; Feng, S.S. Formulation, characterization, and in vitro evaluation of quantum dots loaded in poly(Lactide)-Vitamin E TPGS nanoparticles for cellular and molecular imaging. *Biotechnol. Bioeng.* **2008**, *101*, 622–633. [CrossRef] [PubMed]
4. Song, W.; Zheng, Z.; Lu, H.; Wang, X. Incorporation of multi-walled carbon nanotubes into biodegradable telechelic prepolymers. *Macromol. Chem. Phys.* **2008**, *209*, 315–321. [CrossRef]
5. Wu, C.S.; Liao, H.T. Modification of biodegradable polylactide by silica and wood flour through a sol-gel process. *J. Appl. Polym. Sci.* **2008**, *109*, 2128–2138. [CrossRef]
6. Chaunier, L.; Leroy, E.; Della Valle, G.; Dalgalarrondo, M.; Bakan, B.; Marion, D.; Madec, B.; Lourdin, D. 3D printing of maize protein by fused deposition modeling. *AIP Conf. Proc.* **2017**, *1914*, 190003. [CrossRef]
7. Semba, T.; Kitagawa, K.; Ishiaku, U.S.; Hamada, H. The effect of crosslinking on the mechanical properties of polylactic acid/polycaprolactone blends. *J. Appl. Polym. Sci.* **2006**, *101*, 1816–1825. [CrossRef]
8. Holland, B.J.; Hay, J.N. Thermal degradation of nylon polymers. *Polym. Int.* **2000**, *49*, 943–948. [CrossRef]
9. Liu, G.; Liao, Y.; Ma, X. Thermal behavior of vehicle plastic blends contained acrylonitrile-butadiene-styrene (ABS) in pyrolysis using TG-FTIR. *Waste Manag.* **2017**, *61*, 315–326. [CrossRef] [PubMed]
10. Ellis, D.I.; Broadhurst, D.; Goodacre, R. Rapid and quantitative detection of the microbial spoilage of beef by Fourier transform infrared spectroscopy and machine learning. *Anal. Chim. Acta* **2004**, *514*, 193–201. [CrossRef]
11. Argyri, A.A.; Panagou, E.Z.; Tarantilis, P.A.; Polysiou, M.; Nychas, G.J.E. Rapid qualitative and quantitative detection of beef fillets spoilage based on Fourier transform infrared spectroscopy data and artificial neural networks. *Sens. Actuators B. Chem.* **2010**, *145*, 146–154. [CrossRef]
12. Sattlecker, M.; Baker, R.; Stone, N.; Bessant, C. Support vector machine ensembles for breast cancer type prediction from mid-FTIR micro-calcification spectra. *Chemom. Intell. Lab. Syst.* **2011**, *107*, 363–370. [CrossRef]

13. Hancox, N.L. Thermal effects on polymer matrix composites: Part 2. Thermal degradation. *Mat. Design* **1998**, *19*, 93–97. Available online: https://www.scopus.com/record/display.uri?eid=2-s2.0-0032105161&origin=resultslist&sort=cp-f&src=s&st1=The+degradation+of+polymers+thermoplastics+ambient+temperature+condition+&st2=&sid=d2adf8641efbf71f603a9005d37b2c8f&sot=b&sdt=b&sl=88&s=TITLE-ABS-KEY%28 (accessed on 20 March 2018). [CrossRef]
14. Nair, V.; Hinton, G.E. Rectified Linear Units Improve Restricted Boltzmann Machines. In Proceedings of the 27th International Conference on Machine Learning (ICML-10), Haifa, Israel, 21–24 June 2010; pp. 807–814.
15. Carrasco, F.; Pagès, P.; Gámez-Pérez, J.; Santana, O.O.; Maspoch, M.L. Processing of poly(lactic acid): Characterization of chemical structure, thermal stability and mechanical properties. *Polym. Degrad. Stable.* **2010**, *95*, 116–125. [CrossRef]

![applied sciences logo] *applied sciences*

**MDPI**

*Article*

# A Thermomechanical Analysis of Conformal Cooling Channels in 3D Printed Plastic Injection Molds

Suchana Akter Jahan [1,*] and Hazim El-Mounayri [2]

[1] School of Mechanical Engineering, Purdue University, West Lafayette, IN 47907, USA
[2] Mechanical and Energy Engineering Department, Indiana University Purdue University Indianapolis, 723 W Michigan St, Indianapolis, IN 46202, USA; helmouna@iupui.edu
* Correspondence: sjahan@purdue.edu; Tel.: +1-317-954-4133

Received: 1 November 2018; Accepted: 7 December 2018; Published: 11 December 2018

**Abstract:** Plastic injection molding is a versatile process, and a major part of the present plastic manufacturing industry. The traditional die design is limited to straight (drilled) cooling channels, which don't impart optimal thermal (or thermomechanical) performance. With the advent of additive manufacturing technology, injection molding tools with conformal cooling channels are now possible. However, optimum conformal channels based on thermomechanical performance are not found in the literature. This paper proposes a design methodology to generate optimized design configurations of such channels in plastic injection molds. The design of experiments (DOEs) technique is used to study the effect of the critical design parameters of conformal channels, as well as their cross-section geometries. In addition, designs for the "best" thermomechanical performance are identified. Finally, guidelines for selecting optimum design solutions given the plastic part thickness are provided.

**Keywords:** additive manufacturing; design of experiments; conformal cooling channels; design and analysis; design rules for additive manufacturing

## 1. Introduction

Our everyday life is filled with various types of plastic products. Plastic injection molding is a widely used manufacturing process that consumes a large percentage of the total amount of plastics [1]. Different complex sizes and shapes of high-quality products can be produced by this versatile process from thermoplastic and thermosetting materials with the application of heat and pressure [2]. The design of die core and cavity is very critical in the economic aspects of the injection molding business, a better quality and cost-effective product is of the utmost importance in today's competitive market.

The cooling of injection molding tooling plays a very important role in the total production cycle time of the injection molding process. The cooling time constitutes about half of the overall production cycle; hence, it is a significant portion of the entire molding process [3]. The effect of the process parameters on the polymer crystallization in plastic injection molding can be found in [4].

Injection molding is a highly used multipurpose manufacturing process for the production of plastic parts that is accepted all over the world. Traditionally, there are straight drilled holes in the solid dies that help cool the hot molten plastic inside the cavity. The cooling step is a major portion of the production cycle, resulting in a high cost of production. With the rising competition worldwide in the plastic product business, it is very important to reduce the production cost, which can be achieved by reducing the production cycle time. Implementing the design of the conformal cooling channel is a good choice for this purpose. Conformal cooling channels can improve the performance of the molds in many ways, for example; uniform and fast cooling, less warping and defects, etc. As we know that any kind of channels could be produced using the additive manufacturing process, this technology can bring tremendous development and business benefits to the plastic injection molding industry.

The use of cooling channels that are conformal to the molding cavity can improve the control of the mold temperature and part dimensions, as reported by a group at MIT in the 1990s [5]. Xu and Sachs at MIT presented a modular, systematic approach for the design of conformal cooling channels. They divided the tool into different geometric regions and created the channel systems for each region, recognizing cooling as local to the surface of the tool. The different design parameters that were considered in the study considered in the study included the mold surface temperature, pressure drop, mold material strength, etc. [5–7]. Three-dimensional (3D) printing technology was implemented for the direct fabrication of tooling using metal powders, and improved the thermal management, dimensional control, surface finishing, and tool hardening. Stainless steel powder with a resultant tooling hardness of 25–30 Rockwell C was used in the study [8].

A study on rapid soft tooling for plastic injection molding was conducted by Ferreira and Mateus. They proposed some original approaches to integrate the advanced processing technologies, which featured composite materials chilled by conformal cooling channels in injection molding tools [9]. Meckley and Edwards presented the effect of a conformal cooling channel on reducing the cooling time and increasing part quality, in comparison with traditional straight cooling channels. They used high-density polyethylene and polycarbonate in their study, and demonstrated that the mold and melt temperature differences between the two materials illustrated the efficiency of the conformal channels [10]. The use of conformal cooling channels to both heat and cool a single injection molding tool was demonstrated in the research work by Hopkins and Dickens. This paper discussed the potential of 3D printing technology for achieving the successful production of complex geometries [11]. Yoo provided an investigation on the advantages of rapid tooling methods to build heating and cooling channels in order to enhance thermal control [12]. He also demonstrated how to seal such channels rapidly and inexpensively.

Altaf et al. provided an insight on the conformal channel fabrication method, which is not possible using traditional drilling or machining processes [13]. They presented a technique for fabricating conformal cooling channels in an aluminum-filled epoxy mold using Rapid Prototyping (RP) techniques. An investigation of the automation of the preliminary design stage to the layout design stage of the cooling system design process was presented by Li et al., who provided a configuration of straight cooling channels based on the size and shape of the plastic part design, which does not necessarily require an additive manufacturing technique [14].

There have been a series of studies in the area of design and modeling of conformal cooling channels in injection molding tooling for a long time, yet the concept of simulating the designs cannot be traced back more than 10–12 years. Since then, researchers have been using different commercially available simulation packages to analyze the tool and channel designs. Dima et al. found the best position of the runner in 2005, using Moldflow analysis in I-DEAS™ [3]. Two years later, Saifullah and Masood analyzed part cooling times using the ANSYS thermal analysis [15]. Two more years later, this research group demonstrated comparative part analysis using MPI simulation software for conventional and square section conformal cooling channels; they concluded that conformal channels render 35% less cooling time than conventional ones [16]. By incorporating a square sectioned conformal cooling channel system for injection mold dies, they provided comparative studies between conformal and traditional molds [17]. Finite element analysis using ANSYS was also presented for a mold with bimetallic conformal cooling channels. They compared the performance with a conventional mold, and provided experimental verification with two different plastic materials that were produced by a miniature injection molding machine [16]. Xu and Sachs presented a quantitative guidance for tooling design in 2009. Their proposed methodology was tested on a 3D printed benchmark tool with truss support. In their study, preliminary tests demonstrated the technical feasibility of using a solid freeform fabrication process to create low thermal inertia tools [18].

Another finite element study was presented by Sun et al. for the milled groove insert method for the cooling of plastic injection molds using a household iron plastic part [19]. This analysis was based on a cooling and thermal stress modeling technique [20]. Gloinn et al. performed finite element analysis (FEA) to determine the mold temperature using ABS polymer as the molten material and water at 20 °C as the cooling fluid [21]. Au and Yu conducted a study to investigate the thermal effects of cooling channel design on injection molding using Moldflow Plastic Insight 3.1 [22]. They proposed a novel scaffold for the design of uniform conformal cooling. A few years later in 2013, Hsu et al. identified that for cavities with irregular geometry, the distance between the cooling channels and the cavity would vary throughout the part, and would cause local heat accumulation and product defects such as sink mark, warpage, etc. They adopted a true 3D simulation technique to predict the cooling time and compare the results with traditional molds [23]. Dang and Park adopted an algorithm to calculate the temperature distribution through molds, and presented a conformal channel design with an array of baffles for obtaining uniform cooling over the entire freeform surface of the molded parts [24]. In addition to that, they provided an insight into the use of conformal cooling channels to provide a uniform cooling and reduce the cycle time for the injection molding process. U-shaped milled groove conformal channels were presented, and an optimization process was also proposed to obtain an optimal configuration of the conformal channels [25]. The comparative effect of conventional, series, parallel, and additive parallel cooling channels was studied by Khan et al. in 2014 with respect to cooling time, total cycle time, volumetric shrinkage, and temperature variance using AMI software [26]. In another study, Zink et al. pointed out the effect of limescale on cooling efficiency, also in the cases of conformal cooling channels [27].

Wang et al. presented an automatic method for designing conformal cooling circuits by establishing a relationship between the conformal cooling and the shape of the plastic body [28]. Choi et al. established a higher degree of freedom in the design of conformal cooling channels with the application of additive manufacturing and concentrate on a branching law principle to improve the cooling efficiency in injection molds [29]. To create the design of conformal cooling channels, they used the Voronoi diagram algorithm and the binary branching algorithm. A similar technique was also adopted by Park and Pham. They designed cooling channels for individual surfaces, and then combined them to form an overall conformal cooling channel system for the entire part [30]. Two years later, they designed conformal cooling channels for an automotive part using the algorithm that they provided in their previous work. In that study, they conducted an optimization to minimize the cooling time with boundaries ensuring a realistic design for the cooling system [31]. Wang et al. introduced an approach to generate spiral channels for conformal cooling. Using boundary distance maps, their algorithm could generate evenly distributed spiral channels in the injection mold [32].

In 2011, a design methodology called visibility-based cooling channel generation was presented by Au and Yu for an automatic preliminary cooling channel design. This was more of a geometric and theoretical method, rather than intended for a practical scenario [33]. Subsequently, this research group provided a cooling channel distance modification based on adjustments to the direction and amount in 2014. Also, a simulation technique using MoldFlow Plastic Insight software was adopted to demonstrate the feasibility of their proposed method [34]. Agazzi et al. proposed a new methodology called "Morpho Cooling" for the design of cooling channels in the injection mold. This method provided better results in cooling in terms of the higher uniformity of temperature distribution and lesser part warpage [35].

It is evident that there have been a lot of studies about the analysis of conformal cooling channels, yet, the number of studies dedicated to the design parameters of conformal channels for various kinds of part designs is very limited. In the mold and tooling industry, most of the designs are done based on the designers' experiences. Also, any kind of mix and match between the design parameters, cross-section size, and respective experimental analyses is pretty rare, according to the author's knowledge. Yet, some preliminary information could be gathered from the literature that act as a basis for further research on this project. For example, a simple relationship between four parameters for the design of conformal cooling channels using additive manufacturing is found from Mayer [36]. The data is adapted from their study, and this relationship is shown in Table 1. Some studies show that the use of different cross-sections for channels other than circular might provide better cooling efficiency.

**Table 1.** Correlation amongst design parameters of conformal cooling channels [37].

| Wall Thickness of Molded Part (mm) | Channel Diameter, D (mm) | Pitch Distance, P (mm) | Channel Centerline to Mold Wall Distance, L (mm) |
|---|---|---|---|
| 0–2 | 4–8 | 2D–3D | 1.5D–2D |
| 2–4 | 8–12 | 2D–3D | 1.5D–2D |
| 4–8 | 12–14 | 2D–3D | 1.5D–2D |

From the literature review, it is evident that there have been a number of studies in the field of design that provide an analysis of conformal cooling channels in injection molds. Some of the studies mostly discussed the process of designing conformal channels and the design parameters. Some other papers concentrated on the production process of such conformal cooling channels. Again, some provided results of numerical analyses of injection molds. Yet, no study has yet identified how to incorporate conformal cooling channels into injection molds in order to most effectively provide both thermal and structural performance. This is the principal motivating objective of this research effort. The effort of identifying a design technique for optimal conformal cooling channels in injection molds has been initiated in recent years by Jahan and Wu et al. [37–43]. This current paper provides a guideline for the mold designers to design their injection molds with the conformal cooling channels that would enable them to obtain the most benefit in business. A number of sets of design of experiments (DOEs) are presented, where thermomechanical analysis is performed on all the design cases in all of the DOEs, and the outcomes are compared to finally identify the most effective design configuration for the conformal cooling channels. This optimization is expected to provide useful insight for mold designers in the plastic injection molding industry.

## 2. Materials and Methods

We propose a methodology for creating conformal cooling channels in injection molds that result in better performance regarding fast and uniform cooling, structural stability, reduced cycle time, and improved part quality. By applying additive manufacturing, we can create cooling channels of any size and shape inside the mold core and cavity, which is not possible using traditional manufacturing processes.

In the recent publication by the current authors [37,42], numerical models to analyze the thermal and structural behavior of plastic injection molds were developed and validated. In that study, the authors generated an optimal design of conformal cooling channels for a specific size and shape of a plastic part. Moreover, a cooling channel with a rectangular cross-section provided shorter cooling time. In this paper, the authors propose a general design methodology for conformal cooling channels. Here, three different values of the thickness of plastic parts are considered, and a thermomechanical optimization is achieved through a design of experiments approach. The design of the plastic part to be manufactured by injection molding affects the design of the cooling system of the mold. For example, the higher the thickness of the plastic part, the larger the cooling channel diameter should be. Mayer [36] provided a guideline showing the correlations between the design variables of conformal cooling channels. In the current study, the design variables are kept independent in all of the DOEs, and are considered within a range to allow for a comprehensive design solution.

### 2.1. Design of Experiments for Optimized Conformal Channels

A design of experiments approach is used to guide the design of conformal cooling channels. Typically, the thickness of molded plastic parts varies within a range of 0–6 mm. Different shapes, sizes, and thicknesses of the part would require different configurations of optimized conformal cooling channels. As such, only one basic shape of plastic parts, namely cylindrical, is considered. Three different thicknesses are modeled: one mm, 3.5 mm, and six mm. Thus, three sets of DOEs are prepared: DOE-1 for the thickness of one mm; DOE-2 for the thickness of 3.5 mm; and DOE-3 for the thickness of six mm.

The information regarding the part thickness provides the basic outline for channel design parameters such as diameter (in circular channels), pitch distance, channel centerline to mold wall distance, etc. The range of analysis for each DOE has been determined from the literature and the general rule of thumb of mold designers [36]. After deciding on the design variables (pitch, wall to wall distance, etc.), the channel cross-section is selected. This research group found that a rectangular cross-section of conformal cooling channels provides effective thermal performance in injection molds. As a result, all of the design cases in the above-mentioned DOEs are created with rectangular-shaped cooling channels. The DOE parameter details are mentioned in Table 2. The design cases are designated as 1.1, 1.2 . . . , 2.1 . . . , 3.1, 3.2, . . . 3.18.

**Table 2.** Case setup variables in design of experiments (DOEs) 1, 2 and 3, in which the thickness of the plastic parts are set at one mm, 3.5 mm, and six mm, respectively.

| | DOE-1 | | | | DOE-2 | | | | DOE-3 | | |
|---|---|---|---|---|---|---|---|---|---|---|---|
| Case Number | X-Section (mm × mm) | P (mm) | L (mm) | Case Number | X-Section (mm × mm) | P (mm) | L (mm) | Case Number | X-Section (mm × mm) | P (mm) | L (mm) |
| 1.1 | 3.8 × 2.5 | 8 | 6 | 2.1 | 7.5 × 5 | 16 | 12 | 3.1 | 11.3 × 7.5 | 24 | 18 |
| 1.2 | 3.8 × 2.5 | 8 | 16 | 2.2 | 7.5 × 5 | 16 | 24 | 3.2 | 11.3 × 7.5 | 24 | 28 |
| 1.3 | 3.8 × 2.5 | 16 | 6 | 2.3 | 7.5 × 5 | 26 | 12 | 3.3 | 11.3 × 7.5 | 33 | 18 |
| 1.4 | 3.8 × 2.5 | 16 | 16 | 2.4 | 7.5 × 5 | 26 | 24 | 3.4 | 11.3 × 7.5 | 33 | 28 |
| 1.5 | 3.8 × 2.5 | 24 | 6 | 2.5 | 7.5 × 5 | 36 | 12 | 3.5 | 11.3 × 7.5 | 42 | 18 |
| 1.6 | 3.8 × 2.5 | 24 | 16 | 2.6 | 7.5 × 5 | 36 | 24 | 3.6 | 11.3 × 7.5 | 42 | 28 |
| 1.7 | 5.6 × 3.8 | 8 | 6 | 2.7 | 9.4 × 6.3 | 16 | 12 | 3.7 | 12.3 × 8.2 | 24 | 18 |
| 1.8 | 5.6 × 3.8 | 8 | 16 | 2.8 | 9.4 × 6.3 | 16 | 24 | 3.8 | 12.3 × 8.2 | 24 | 28 |
| 1.9 | 5.6 × 3.8 | 16 | 6 | 2.9 | 9.4 × 6.3 | 26 | 12 | 3.9 | 12.3 × 8.2 | 33 | 18 |
| 1.10 | 5.6 × 3.8 | 16 | 16 | 2.10 | 9.4 × 6.3 | 26 | 24 | 3.10 | 12.3 × 8.2 | 33 | 28 |
| 1.11 | 5.6 × 3.8 | 24 | 6 | 2.11 | 9.4 × 6.3 | 36 | 12 | 3.11 | 12.3 × 8.2 | 42 | 18 |
| 1.12 | 5.6 × 3.8 | 24 | 16 | 2.12 | 9.4 × 6.3 | 36 | 24 | 3.12 | 12.3 × 8.2 | 42 | 28 |
| 1.13 | 7.5 × 5 | 8 | 6 | 2.13 | 11.3 × 7.5 | 16 | 12 | 3.13 | 13.2 × 8.8 | 24 | 18 |
| 1.14 | 7.5 × 5 | 8 | 16 | 2.14 | 11.3 × 7.5 | 16 | 24 | 3.14 | 13.2 × 8.8 | 24 | 28 |
| 1.15 | 7.5 × 5 | 16 | 6 | 2.15 | 11.3 × 7.5 | 26 | 12 | 3.15 | 13.2 × 8.8 | 33 | 18 |
| 1.16 | 7.5 × 5 | 16 | 16 | 2.16 | 11.3 × 7.5 | 26 | 24 | 3.16 | 13.2 × 8.8 | 33 | 28 |
| 1.17 | 7.5 × 5 | 24 | 6 | 2.17 | 11.3 × 7.5 | 36 | 12 | 3.17 | 13.2 × 8.8 | 42 | 18 |
| 1.18 | 7.5 × 5 | 24 | 16 | 2.18 | 11.3 × 7.5 | 36 | 24 | 3.18 | 13.2 × 8.8 | 42 | 28 |

For DOE-1, the plastic part thickness is one mm. According to the guidelines in Table 1, the channel diameter should be in the range of four to eight mm. The perimeter of such channels would be 12.6 mm to 25.13 mm. Keeping the perimeter the same, the circular channels are converted into rectangular ones, and their cross-sectional dimensions are calculated. For example, the circular channel with a diameter of four-mm was converted to a rectangular channel with a cross-section of 3.8 mm × 2.5 mm, and the channel with a diameter of eight mm is converted into a 7.5 mm × 5 mm section channel. There are three design variables in each DOE 1, 2 and 3. These are the channel cross-section (a × b), pitch distance (P), and mold wall to channel centerline distance (L). The first two variables have three levels of design, and the third one has two levels of design. As a result, DOE-1 has 3 × 3 × 2 = 18 design cases (Table 2, column 1). Similarly, DOE-2 has 18 cases (Table 1, column 5), and DOE-3 also has 18 design cases (Table 2, column 9). The channel design parameters, such as pitch distance and cross-section dimensions, are the same for both the cavity and core in a single design case. In all of the DOEs mentioned here (DOE 1, 2 and 3), as well as the other ones (DOE 4, 5, 6, 7, 8 and 9) discussed in the latter part of the paper, a full factorial method is used to formulate the design of experiments.

## 2.2. Thermomechanical Optimization

In previous studies by this research group [37–45], the numerical analysis and design decisions regarding the performance of conformal cooling channels in injection molds were conducted with the single consideration of fast cooling. In addition, static structural analysis was conducted to ensure the structural stability of the mold cores and cavities with conformal channels. Both thermal and structural analyses were performed on all of the design case studies, and a trade-off between thermal and structural performance was conducted to find the best design solution.

Theoretically, a solid mold with a single straight-drilled channel is structurally stable, and can withstand large amount of stress compared to a mold with conformal channels due to the higher void space within the body. The mold is made of structural steel (density: 7850 kg/m$^3$, thermal conductivity 60.5 W/m-K, specific heat 60.5 J/kg-K, yield strength 430 MPa). On the other hand, a mold with conformal channels can cool off the molten plastic quickly, as well as be uniformly compared to the traditional mold due to the presence of conformal channels at the vicinity of the plastic part. The plastic is polypropylene (density: 830 kg/m$^3$, thermal conductivity 0.14 W/m-K, specific heat 1900 J/kg-K). In the thermal analysis, the initial condition was set for coolant inlet at 22 °C, and the molten plastic was set at 168 °C. The plastic ejection temperature was 50 °C. For the mechanical or static structural analysis, the clamping force was 110 ton, which was applied to the top and bottom surface of the mold. In addition to that, an injection pressure of 131 MPa was applied to the heating surface. The structural analysis predicted the deformation and distribution of von Mises stress on the mold body. If the maximum von Mises stress is below the acceptable limit of yield stress of the mold material, the mold is considered to be structurally stable and functional. The design cases as mentioned in Table 2 are analyzed for both the thermal and structural behavior using the simulation technique in the published research of the author [38,42], and an optimization method is conducted for each DOE to obtain the best design scenario. The optimization problem statements for DOE-1, DOE-2, and DOE-3 are summarized in Table 3.

**Table 3.** Optimization problem statements for DOE-1, DOE-2, and DOE-3.

|  | DOE-1 | DOE-2 | DOE-3 |
|---|---|---|---|
| Objective function | 1. Minimize cooling time | 1. Minimize cooling time | 1. Minimize cooling time |
|  | 2. Minimize max von Mises stress | 2. Minimize max von Mises stress | 2. Minimize max von Mises stress |
| Design Variables | • Channel perimeter (C) | • Channel perimeter (C) | • Channel perimeter (C) |
|  | • Pitch (P) | • Pitch (P) | • Pitch (P) |
|  | • Mold wall to channel centerline distance (L) | • Mold wall to channel centerline distance (L) | • Mold wall to channel centerline distance (L) |
| Constraints | • Cooling time <28.04 s | • Cooling time <28.32 s | • Cooling time <35.55 s |
|  | • Maximum von Mises stress <215 MPa | • Maximum von Mises stress <215 MPa | • Maximum von Mises stress <215 MPa |
|  | • 12 mm< C <25 mm | • 25 mm< C <38 mm | • 382 mm< C <44 mm |
|  | • 8 mm< P <24 mm | • 16 mm< P <36 mm | • 24 mm< P <42 mm |
|  | • 16 mm< L <26 mm | • 12 mm< L <24 mm | • 18 mm< L <28 mm |

*2.3. Design Optimization of Cooling Channels for Different Shape of Plastic Parts*

The previously mentioned DOEs (DOE 1, 2 and 3) were designed considering that the final plastic part is only cylindrical in shape. However, in the real world, there are a number of various sizes and shapes of plastic products. To study the effect of this variation in shape and how it affects the optimal shape of the conformal cooling channels, six new sets of DOEs are prepared (DOE 4, 5, 6, 7, 8 and 9). Six different plastic part designs are considered here. The CAD models of these plastic parts are shown in Figure 1. There are two basic part shapes, each with three different thicknesses. Hence, six different design of experiments sets are created. Table 4 shows the DOE set with the plastic part dimension.

The design of the conformal cooling channels for these DOEs are based on the results obtained from DOE 1, 2 and 3. The best possible design solutions found from DOE 1, 2 and 3 are used as design baselines for the ones in DOE 4, 5, 6, 7, 8 and 9. This is explained in detail in the later sections.

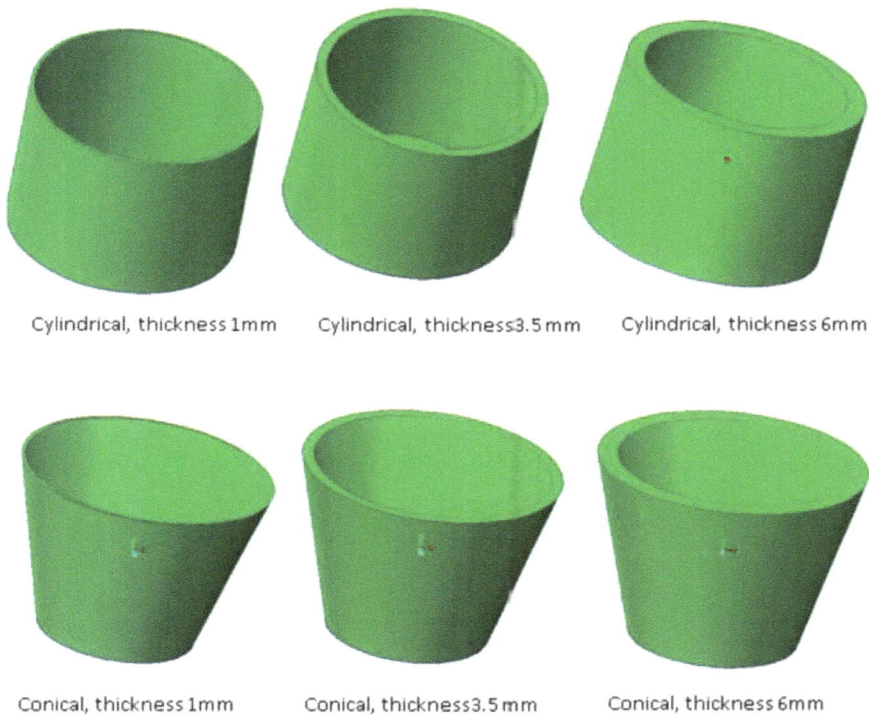

Figure 1. CAD models of six different plastic parts.

**Table 4.** Plastic part dimensions for DOE 4, 5, 6, 7, 8 and 9.

| DOE Number | Shape of Plastic Part | Part Thickness (mm) | Part Height (mm) | Part Larger Diameter (mm) | Part Smaller Diameter (mm) |
|---|---|---|---|---|---|
| DOE-4 | cylindrical | 1 | 60 | 80 | 80 |
| DOE-5 | cylindrical | 3.5 | 60 | 80 | 80 |
| DOE-6 | cylindrical | 6 | 60 | 80 | 80 |
| DOE-7 | conical | 1 | 60 | 80 | 50 |
| DOE-8 | conical | 3.5 | 60 | 80 | 50 |
| DOE-9 | conical | 6 | 60 | 80 | 50 |

## 3. Results and Discussions

### 3.1. Findings from DOE 1, 2 and 3

Table 5 shows the thermal and structural results obtained from DOE 1, 2 and 3. The results are in terms of cooling time and maximum von Mises stress at the time of ejection. The objective is to obtain the minimum cooling time and minimum stress, as previously mentioned in Table 3.

In Table 5 column 2, it is seen that for DOE-1, the minimum cooling time is 14.32 s in case 1.1, whereas the minimum value of the maximum von Mises stress occurs in case 1.10 (column 3, row 12), which is 107 MPa. Similar phenomena are observed for DOE-2 and DOE-3, too. In DOE-2 (Table 5 column 5), the minimum cooling time, i.e., 21.39 s, occurs in case 2.7, whereas the minimum stress 161 MPa occurs in case 2.5 (column 6). In DOE-3, case 3.13 and case 3.6 have the best thermal (27.47 s) and best structural (162 MPa) results, as found in column 8 and 9, respectively. Hence, a trade-off is necessary in order to define the best design solution.

**Table 5.** DOE 1, 2 and 3 thermal and structural results.

| DOE-1 | | | DOE-2 | | | DOE-3 | | |
|---|---|---|---|---|---|---|---|---|
| Case Number | Cooling Time (s) | Max von Mises Stress (MPa) | Case Number | Cooling Time (s) | Max von Mises Stress (MPa) | Case Number | Cooling Time (s) | Max von Mises Stress (MPa) |
| 1.1 | 14.32 | 159 | 2.1 | 24.01 | 183 | 3.1 | 28.78 | 179 |
| 1.2 | 23.83 | 155 | 2.2 | 34.59 | 176 | 3.2 | 31.91 | 180 |
| 1.3 | 14.68 | 157 | 2.3 | 28.57 | 169 | 3.3 | 32.52 | 173 |
| 1.4 | 35.97 | 206 | 2.4 | 39.22 | 164 | 3.4 | 35.39 | 170 |
| 1.5 | 26.07 | 188 | 2.5 | 32.63 | 161 | 3.5 | 35.04 | 172 |
| 1.6 | 33.36 | 156 | 2.6 | 42.87 | 162 | 3.6 | 37.15 | 162 |
| 1.7 | 17.97 | 165 | 2.7 | 22.37 | 203 | 3.7 | 28.35 | 197 |
| 1.8 | 20.08 | 299 | 2.8 | 32.59 | 196 | 3.8 | 30.78 | 174 |
| 1.9 | 17.53 | 159 | 2.9 | 26.55 | 166 | 3.9 | 36.6 | 184 |
| 1.10 | 25.22 | 107 | 2.10 | 37.33 | 163 | 3.10 | 34.24 | 168 |
| 1.11 | 23.04 | 153 | 2.11 | 31.37 | 215 | 3.11 | 33.99 | 174 |
| 1.12 | 29.64 | 161 | 2.12 | 40.71 | 168 | 3.12 | 36.35 | 171 |
| 1.13 | N/A | N/A | 2.13 | 23.58 | 244 | 3.13 | 27.47 | 176 |
| 1.14 | N/A | N/A | 2.14 | 33.45 | 222 | 3.14 | 30.52 | 189 |
| 1.15 | 16.49 | 174 | 2.15 | 26.75 | 171 | 3.15 | 30.75 | 186 |
| 1.16 | 23.88 | 204 | 2.16 | 36.03 | 169 | 3.16 | 33.35 | 177 |
| 1.17 | 21.21 | 144 | 2.17 | 29.98 | 165 | 3.17 | 33.45 | 177 |
| 1.18 | 28.1 | 169 | 2.18 | 38.57 | 167 | 3.18 | 35.65 | 175 |

It is notable that the value of the minimum cooling time increases from DOE-1 to DOE-3. This is an expected behavior, as the thickness of the plastic part also increases. Although the cooling time increases, in each case, they show better results than their respective traditional mold design scenario. This comparison is conducted by creating three traditional mold designs with straight drilled cooling channels and analyzing their thermal and structural performance. These cases are named as 1-conventional, 2-conventional, and 3-conventional. These design cases are created for plastic parts with thicknesses of one mm, 3.5 mm, and six mm, respectively, in order to be comparable with their conformal design cases. Table 6 shows the cooling times for these conventional cases along with the respective conformal design cases, which indicate positive improvement with the application of conformal cooling channels in all of the cases.

**Table 6.** Comparative thermal and structural results from DOE 1, 2 and 3 with conformal and conventional cooling channels.

| DOE Number | Conformal Cooling Time (s) | Conventional Cooling Time (s) | Conformal Max. von Mises Stress (MPa) | Conventional Max. von Mises Stress (MPa) |
|---|---|---|---|---|
| DOE-1 | 14.32 | 28.04 | 159 | 153 |
| DOE-2 | 22.37 | 28.32 | 203 | 167 |
| DOE-3 | 27.47 | 35.55 | 176 | 168 |

The thermal analysis results for the design cases mentioned in Table 6 are shown in Figure 2. It shows the temperature distribution of the respective plastic parts at the time of ejection. The selected top three design cases are listed in Table 7. This selection is based on a trade-off between the best thermal results and best structural results in DOE 1, 2 and 3.

**Table 7.** Selected top three optimal design cases for DOE 1, 2 and 3.

| DOE Number | Selected Cases | | |
|---|---|---|---|
| DOE-1 | 1.1 | 1.3 | 1.9 |
| DOE-2 | 2.1 | 2.9 | 2.17 |
| DOE-3 | 3.1 | 3.8 | 3.13 |

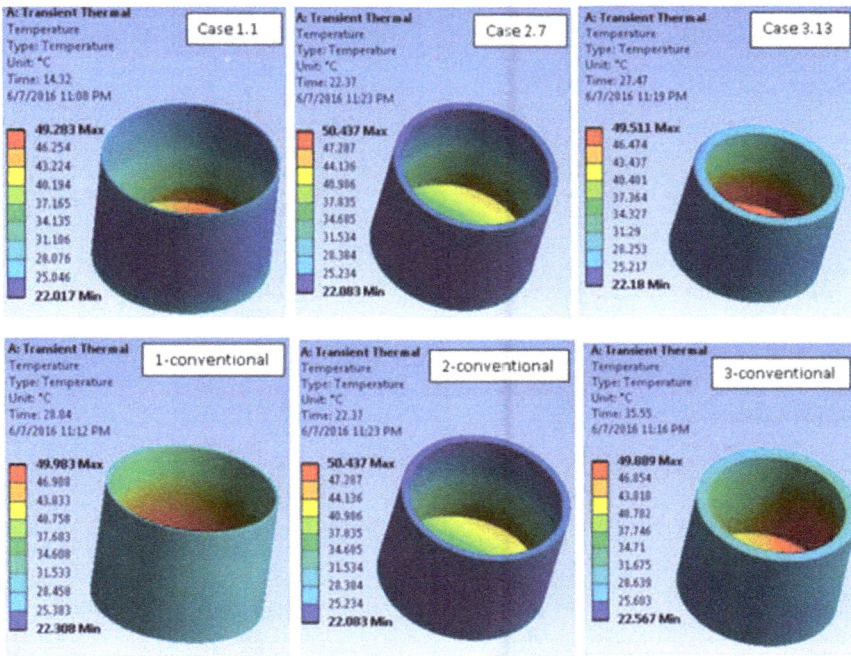

**Figure 2.** Temperature distribution on plastic parts for comparable conformal and conventional cases in DOE 1, 2 and 3.

Figure 3 shows the variation of cooling time and stress distribution with pitch distance (P) and the mold-to-channel centerline distance (L) for DOE-1. Figures 4 and 5 show similar trends for DOE-2 and DOE-3 (i.e., increase in cooling time with increase in P value). The same effect is observed for L. From these figures, it can be noted that with the increase in part thickness, the variation in cooling times decreases. Hence, the thicker the plastic part, the more critical the need for an improved conformal cooling channel design. When the plastic parts are thicker, the temperature is higher in the inner surface than the outer surface, and it is more important to keep the distance of the channels to the mold wall smaller in the core than the cavity.

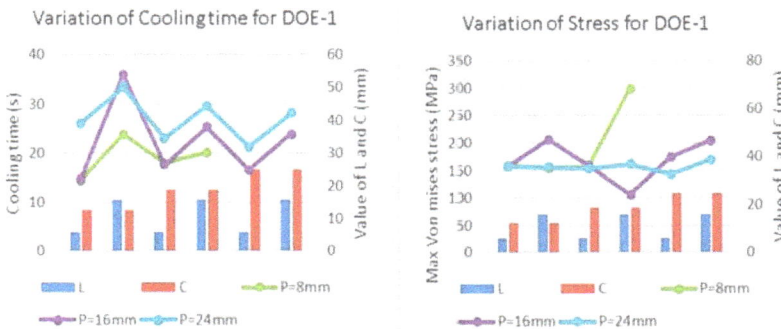

**Figure 3.** Trend of cooling time and maximum von Mises stress variation in DOE-1.

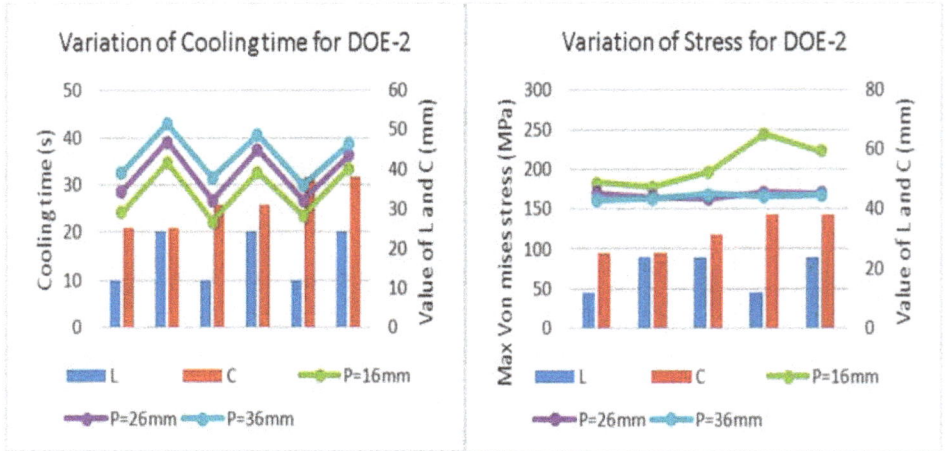

**Figure 4.** Trend of cooling time and maximum von Mises stress variation in DOE-2.

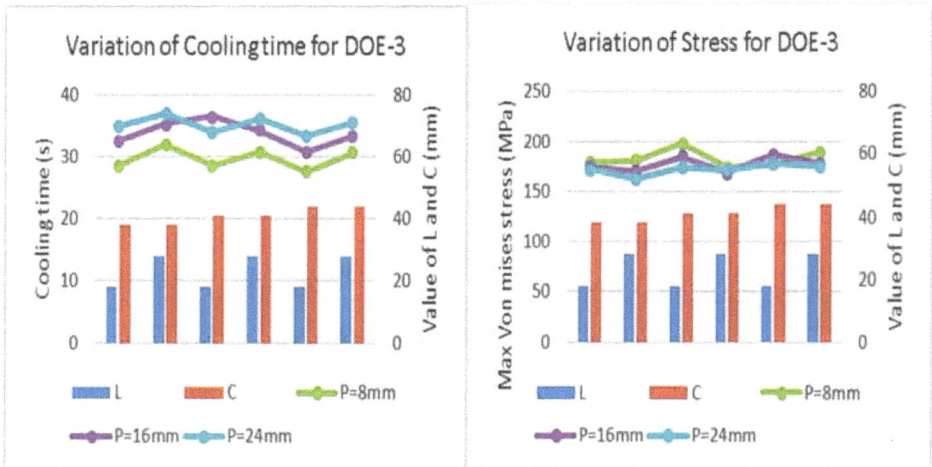

**Figure 5.** Trend of cooling time and maximum von Mises stress variation in DOE-3.

Analyzing the thermal and structural results, it is noted that although the cooling times vary significantly among the cases in each DOE set, the value of the maximum von Mises stress is relatively constant. Those values are much smaller than the yield strength. Thus, the thermal performance is emphasized to determine optimum designs. In DOE 1, 2 and 3, considering the cooling times, the variances are 38.45 s, 35.72 s, and 8.27 s respectively, while the variances are 1576.56 MPa, 563.89 MPa, and 63.43 MPa, considering the von Mises stress. With the optimal designs obtained in the previous section, the study is further extended by considering the cylindrical and conical designs of plastic parts, which are the most commonly used in the injection molding industry. The top three optimized designs of cooling channels (from Table 7) for each thickness are chosen to set up new sets of DOEs. In terms of the channel cross-section, circular, square, and rectangular designs provide comparatively better thermal performance according to a previous study [38]. These three different channel cross-sections are incorporated here.

## 3.2. Setting up DOE 4, 5, 6, 7, 8 and 9 Using Results of DOE 1, 2 and 3

As mentioned in Section 2.3, the setup of DOEs 4, 5, 6, 7, 8, and 9 is completely dependent upon the results of DOE 1, 2 and 3. As we have discussed the results of the first three DOEs already, we can now explain how the DOE 4, 5, 6, 7, 8 and 9 are created. Let's consider Table 7. It contains the optimally selected top results of DOE 1, 2 and 3. In the first row, it shows the top three design cases for DOE-1, which are case 1.1, 1.3, and 1.9. If we go back to Table 2 to understand what the channel design in case 1.1 actually is, we can see that the channel cross-section is 3.8 mm × 2.5 mm, P is 8 mm, and L is 6 mm. Now, to prepare DOE-4, we would like to keep the values of P (pitch) and L exactly the same, and change the shape of the channel cross-section. We will keep the perimeter (circumference) the same and convert the channel design into circular and square-shaped cross-sections. The resulting channels are a circular section with a diameter of four mm and a square section channel of 3.1 mm × 3.1 mm. The values of P and L are the same in all three cases. In this manner, one design case selected in Table 7 provides three different configurations (circular, rectangular, square cross-section) of conformal cooling channel designs. Thus, the nine design cases mentioned in Table 7 provide 9 × 3 = 27 types of channel designs.

These 27 design configurations of conformal cooling channels are incorporated for cylindrical and conical shapes of plastic parts, as earlier mentioned in Section 2.3. Thus, we will have nine design cases the of DOE-4, nine for DOE-5, and so on. DOE 4, 5 and 6 will have a cylindrical plastic part, and DOE 7, 8 and 9 will have a conical plastic part.

Table 8 shows the channel dimensions of these 27 conformal channel designs for cylindrical-shaped plastic parts. These 27 designs can be repeated in injection molds for conical shape plastic parts, which are assigned in DOE 7, 8 and 9. The only difference between the DOE 4, 5 and 6 and DOE 7, 8 and 9 cases is the profile of the channels. For the first case (cylindrical), the channel spiral shape is cylindrical, while the conical-shaped plastic part will have conical-shaped spirals. Hence, the channel dimensions for DOE-4 (Table 8, columns 2, 3, and 4) are exactly the same as those for DOE-7. Similarly, DOE-5 matches with DOE-8, and DOE-6 matches with DOE-9. Figure 6 illustrates the shapes of the conformal channels for cylindrical and conical parts.

**Table 8.** Channel dimensions for DOE 4 (plastic part thickness: one mm), DOE-5 (plastic part thickness: 3.5 mm) and DOE-6 (plastic part thickness: six mm).

| | DOE-4 | | | | DOE-5 | | | | DOE-6 | | |
| --- | --- | --- | --- | --- | --- | --- | --- | --- | --- | --- | --- |
| Case Number | X-Section (mm) | P (mm) | L (mm) | Case Number | X-Section (mm) | P (mm) | L (mm) | Case Number | X-Section (mm) | P (mm) | L (mm) |
| 4.1 | D = 4 | 8 | 6 | 5.1 | D = 8 | 16 | 12 | 6.1 | D = 12 | 24 | 18 |
| 4.2 | 3.8 × 2.5 | 8 | 6 | 5.2 | 7.5 × 5 | 16 | 12 | 6.2 | 11.3 × 7.5 | 24 | 18 |
| 4.3 | 3.1 × 3.1 | 8 | 6 | 5.3 | 6.3 × 6.3 | 16 | 12 | 6.3 | 9.4 × 9.4 | 24 | 18 |
| 4.4 | D = 4 | 16 | 6 | 5.4 | D = 10 | 26 | 12 | 6.4 | D = 13 | 28 | 28 |
| 4.5 | 3.8 × 2.5 | 16 | 6 | 5.5 | 9.4 × 6.3 | 26 | 12 | 6.5 | 12.3 × 8.2 | 28 | 28 |
| 4.6 | 3.1 × 3.1 | 16 | 6 | 5.6 | 7.9 × 7.9 | 26 | 12 | 6.6 | 10.2 × 10.2 | 28 | 28 |
| 4.7 | D = 6 | 16 | 6 | 5.7 | D = 12 | 36 | 12 | 6.7 | D = 14 | 24 | 18 |
| 4.8 | 5.6 × 3.8 | 16 | 6 | 5.8 | 11.3 × 7.5 | 36 | 12 | 6.8 | 13.2 × 8.8 | 24 | 18 |
| 4.9 | 4.7 × 4.7 | 16 | 6 | 5.9 | 9.4 × 9.4 | 36 | 12 | 6.9 | 10.9 × 10.9 | 24 | 18 |

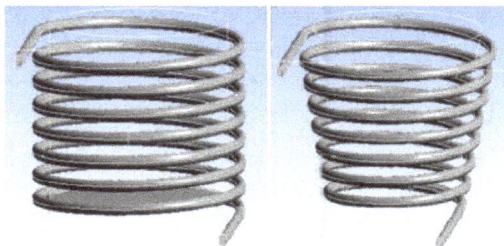

**Figure 6.** Generic cylindrical and conical configuration of conformal cooling channels.

Applying the above-mentioned channel design techniques, six design of experiment sets (DOE 4, 5, 6, 7, 8 and 9) are created and analyzed for their thermal performance. Structural analysis is not conducted for the cases of DOE 4, 5, 6, 7, 8 and 9, as the structural stability has already been checked through simulation.

### 3.3. Findings from DOE1, 2 and 3

Each of the six experimental set designs for DOE 4, 5, 6, 7, 8 and 9 has nine cases. The thermal analysis results in terms of the cooling times of these cases are shown in Table 9. The first six columns are for cylindrical cases, and the later ones are for conical cases.

**Table 9.** Thermal analysis results (cooling times) for DOE 4, 5, 6, 7, 8 and 9

| Cylindrical Part | | | | | | Conical Part | | | | | |
|---|---|---|---|---|---|---|---|---|---|---|---|
| Case Number | Cooling Time (s) | Case Number | Cooling Time (s) | Case Number | Cooling Time (s) | Case Number | Cooling Time (s) | Case Number | Cooling Time (s) | Case Number | Cooling Time (s) |
| 4.1 | 13.51 | 5.1 | 28.82 | 6.1 | 27.65 | 7.1 | 12.7 | 8.1 | 22.63 | 9.1 | 17.63 |
| 4.2 | 14.32 | 5.2 | 24.01 | 6.2 | 28.48 | 7.2 | 12.84 | 8.2 | 19.84 | 9.2 | 17.87 |
| 4.3 | 13.87 | 5.3 | 23.73 | 6.3 | 28.72 | 7.3 | 12.75 | 8.3 | 19.65 | 9.3 | 20.05 |
| 4.4 | 20.03 | 5.4 | 27.50 | 6.4 | 30.35 | 7.4 | 18.03 | 8.4 | 21.48 | 9.4 | 18.13 |
| 4.5 | 14.68 | 5.5 | 26.55 | 6.5 | 30.78 | 7.5 | 18.27 | 8.5 | 21.39 | 9.5 | 17.74 |
| 4.6 | 20.39 | 5.6 | 27.45 | 6.6 | 30.16 | 7.6 | 18.25 | 8.6 | 20.85 | 9.6 | 17.50 |
| 4.7 | 17.46 | 5.7 | 29.30 | 6.7 | 26.89 | 7.7 | 15.94 | 8.7 | 21.02 | 9.7 | 15.96 |
| 4.8 | 17.53 | 5.8 | 29.98 | 6.8 | 27.47 | 7.8 | 16.21 | 8.8 | 22.63 | 9.8 | 16.51 |
| 4.9 | 17.86 | 5.9 | 30.28 | 6.9 | 27.16 | 7.9 | 16.04 | 8.9 | 22.63 | 9.9 | 16.37 |

From Table 9, we can see that when all the other design variables are kept unchanged, the cooling time does vary if the cross-section geometry changes. This means that the effect of the cross-section is important when designing conformal cooling channels for injection molds. From the results, it is also be seen that no single cross-section geometry provides the best performance amongst all of the scenarios.

For DOE-4, the minimum cooling time obtained is 13.51 s, which occurs in case 4.1. This is a circular-shaped conformal channel with D = four mm, P = eight mm, and L = six mm. The design cases 4.2 and 4.3 have the exact same P and L values, but rectangular and square sections provide slightly higher cooling times of 14.32 s and 13.87 s, respectively. The other design cases in DOE-4 provide cooling times in the range of 14 s to 20 s. It is reasonable to mention here that this DOE-4 is only for cylindrical shapes that have one-mm thick plastic parts only, whereas DOE-5 and DOE-6 have 3.5-mm and six-mm plastic parts, respectively. As mentioned earlier, a cylindrical-shaped plastic part with one-mm thickness is cooled down to its ejection temperature at around 28 s. Hence, all of the design cases in DOE-4 (4.1–4.9) are acceptable design solutions, while a mold designer needs to do design one. It is possible for a mold designer to choose and even mix and match from any of these cases if there are some design constraints that do not permit him to choose the most effective one (case 4.1 in this scenario).

For DOE-5, the best result, 23.73 s, is achieved in case 5.3, which is a square section channel (6.3 mm × 6.3 mm), with P = 16 mm and L = 12 mm. The similar design cases (same P and L as case 5.3) provide higher cooling times. Also, the other design cases in DOE-5 provide cooling times between 25–30 s. Again, for DOE-6 (six-mm thickness plastic part), the best solution comes in case 6.7, which is 26.89 s. The channel dimensions in this case are D = 14 mm, P = 24 mm, and L = 18 mm. The overall cooling time range in DOE-6 is 27 s–30 s. Hence, it is clear that a longer time is required to cool the thick plastic parts compared to the thinner ones. All of these design cases provide significant improvement from the traditional straight-drilled designs of injection molds. These also provide insight into a range of variables in which the mold designers can work.

It is notable here that, starting from DOE-4, the variation of cooling time decreases in DOE-5, and then again in DOE-6. This indicates that when designing a conformal cooling channel configuration for a thick plastic part, a very small change can cause an improvement in cooling time. While designing such channels, mold designers need to conduct detailed design of experiments with a wide range of design variables and a small gap between levels to find out the most effective design for their purpose.

On the other hand, when designing for comparatively thin plastic parts, it is easier to find the most effective configuration, as a small change causes a noteworthy difference in cooling time, and designers can work on a specific range of design variables.

DOE-7, DOE-8, and DOE-9 are designed for conical=shaped plastic parts. The best results are 12.70 s, 19.65 s, and 15.96 s, which are obtained in cases 7.1, 8.3, and 9.7, respectively. It is notable here that these cooling times are quite shorter than those in the cylindrical case. As such, designers may consider replacing cylindrical bodies, such as bottle caps, bowls, plastic containers, etc., with conical-shaped ones whenever possible. Moreover, using draft angle in the part design can also help reduce the cooling times, too. DOEs 4–9 show significantly lower values of variances compared to the previous DOEs (1, 2, and 3). As we have considered cooling times only, the variances for DOE 4–9 are 6.13 s, 5.13 s, 1.94 s, 4.98 s, 1.16 s, and 1.30 s. This indicates that the datasets are closer to each other in the latter DOEs compared to the previous ones, hence the rationale behind planning DOEs 4–9 on the basis of whether DOEs 1–3 are correct.

It should be noted that the channel design configuration that provides the minimum cooling time for a one-mm thick cylindrical plastic part is exactly the same for the one-mm thick conical-shaped part, too. This happens again for the 3.5 mm and six-mm thick parts also, indicating that the results are quite reliable, and can be adopted for other shapes of plastic parts such as rectangular boxes or spherical balls. Figures 7–9 show the cooling time trends in different design cases. They provide a comparison of cooling times for similar cylindrical and conical shape designs for plastic bodies. For the same thickness, and with the same channel design, the conical parts are cooled down faster than the cylindrical designs. In Figure 10, the temperature distribution of the plastic parts for the best design cases as obtained in DOE 4, 5, 6, 7, 8 and 9 are shown.

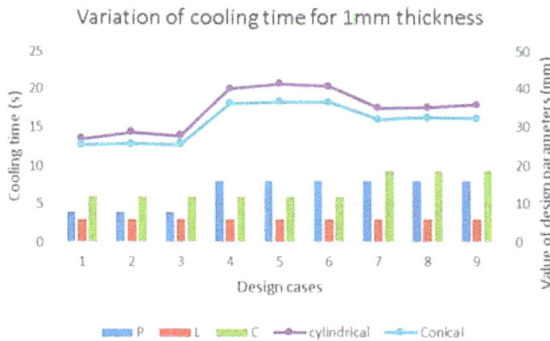

**Figure 7.** Cooling time variation for one-mm thick plastic parts (both cylindrical and conical).

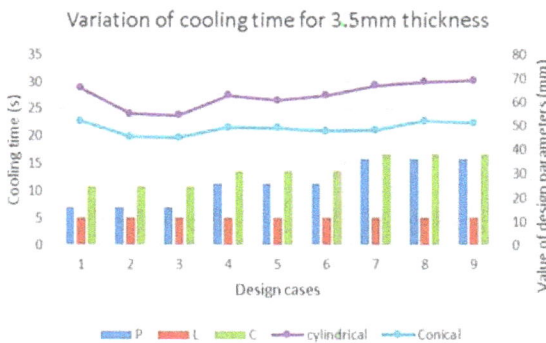

**Figure 8.** Cooling time variation for 3.5-mm thick plastic parts (both cylindrical and conical).

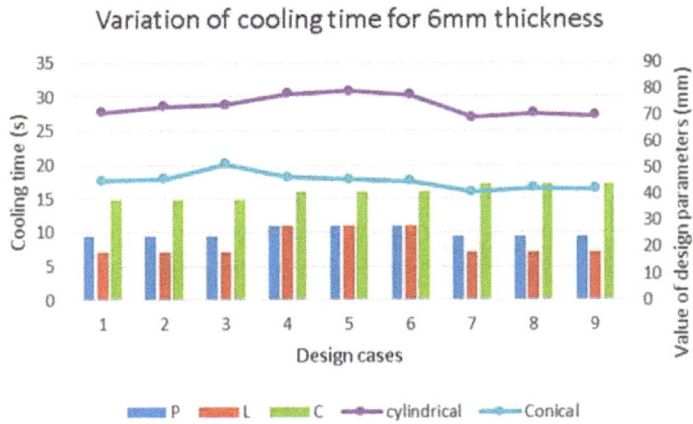

**Figure 9.** Cooling time variation for six-mm thick plastic parts (both cylindrical and conical).

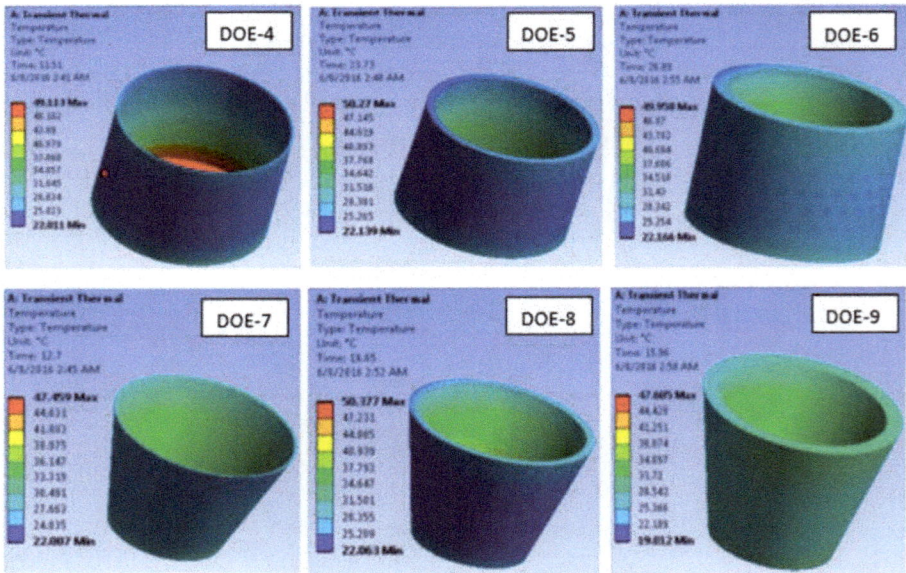

**Figure 10.** Temperature distribution on plastic parts for optimal design cases in DOE 4, 5, 6, 7, 8 and 9.

This study provides a basis and guidelines for mold makers to support their design conformal cooling channels for their injection molds. Due to design constraints in injection molds, such as the position of inlet and outlet ports, the gate, runner, ejector pins, etc., the best design solution may not be feasible in practical industrial cases. For this reason, the top three design cases are enlisted in Figure 11 for each size and shape of plastic parts, in order to provide more options to mold designers in the industry.

| Plastic Part Design | Conformal Channel Design Parameters | | |
| --- | --- | --- | --- |
| | Design Solution 1 (optimum) | Design Solution 2 (less optimal than Solution 1) | Design Solution 3 (less optimal than Solution 2) |
| Cylindrical 1mm | X section: circular, D=4mm, P=8mm, L=6mm | X section: square, 3.1 x3.1(mm x mm), P=8mm, L=6mm | X section: rectangular, 3.8 x2.5(mm x mm), P=8mm, L=6mm |
| Cylindrical 1.5mm | X section: rectangular, 5.6 x3.8(mm x mm), P=8mm, L=7mm | X section: circular, D=6mm, P=8mm, L=7mm | X section: square, 4.7 x4.7(mm x mm), P=8mm, L=7mm |
| Cylindrical 3.5mm | X section: square, 6.3 x6.3(mm x mm), P=16mm, L=12mm | X section: circular, D=8mm, P=16mm, L=12mm | X section: rectangular, 7.5 x5(mm x mm), P=16mm, L=12mm |
| Cylindrical 6mm | X section: circular, D=14mm, P=24mm, L=18mm | X section: rectangular, 13.2x8.8(mm x mm), P=24mm, L=18mm | X section: circular, D=12mm, P=24mm, L=18mm |
| Conical 1mm | X section: circular, D=4mm, P=8mm, L=6mm | X section: square, 3.1 x3.1(mm x mm), P=8mm, L=6mm | X section: rectangular, 3.8 x2.5(mm x mm), P=8mm, L=6mm |
| Conical 3.5mm | X section: square, 6.3 x6.3(mm x mm), P=16mm, L=12mm | X section: rectangular, 7.5 x5(mm x mm), P=16mm, L=12mm | X section: square, 7.9 x7.9(mm x mm), P=26mm, L=12mm |
| Conical 6mm | X section: circular, D=14mm, P=24mm, L=18mm | X section: square, 10.9 x10.9(mm x mm), P=24mm, L=18mm | X section: rectangular, 13.2x8.8(mm x mm), P=24mm, L=18mm |

**Figure 11.** Conformal cooling channel design guidelines.

In addition to the structural stability of the mold and cooling time and production cycle time of the process, the plastic part quality is also important for the plastic injection molding industry. Hence, warping in the final plastic part is a crucial factor to determine the design of the cooling channels in the mold. A detailed study on the effect of warping can be found in the literature [46]. As discussed earlier, the traditional molds have straight drilled holes and buffers in some cases to enhance the cooling inside the mold, which cannot reach very close to the mold walls. On the contrary, "conformal" cooling channels 'conform' to the shape of the mold wall, and are capable of being very near the wall. As a result, the cooling becomes very uniform with such molds, and the warping also minimizes. In this study, warping was not included in the design consideration, as it would elongate the DOEs beyond its scope. In the future, we would incorporate the warping and demonstrate new DOE cases with the study.

## 4. Conclusions

In this study, a design methodology has been implemented to determine the optimum design of conformal cooling channels in injection molds. With the increase of competition in business, mold makers need efficient design tools to serve their purpose. This work provides guidelines to support such a need. A numerical model is developed to analyze the thermal and structural performance of injection molding tools. This model provides a base to compare the performance of various mold designs and conformal channel configurations. A number of design of experiments have been undertaken to identify the most suitable design of channels in injection molds. The results show that for different plastic part designs, different channel configurations provide optimum solutions in terms of the cross-section dimensions, section size, pitch distance, mold wall to channel centerline distance, etc. Hence, a guideline chart is provided in this study to help the mold designers choose the design parameters for their respective cases.

**Author Contributions:** This research article is a combined effort of S.A.J. and her research advisor H.E.-M. Conceptualization, supervision, funding acquisition, review and editing in contributed by H.E.-M., and methodology, software, validation, formal analysis, investigation, resources, data curation, writing, original draft preparation, visualization by S.A.J.

**Funding:** This research was funded by The Walmart Foundation.

*Appl. Sci.* **2018**, *8*, 2567

**Acknowledgments:** The Walmart Foundation supported this research effort through the Walmart U.S. Manufacturing Innovation Fund. Any opinions, findings, conclusions, and recommendations expressed in this investigation are those of the writers and do not necessarily reflect the views of the sponsors.

**Conflicts of Interest:** The authors declare no conflict of interest. The funders had no role in the design of the study; in the collection, analyses, or interpretation of data; in the writing of the manuscript, or in the decision to publish the results.

## References

1. Rosato, D.V.; Rosato, M.G. *Injection Molding Handbook*; Springer Science & Business Media: Berlin, Germany, 2012.
2. Zheng, R.; Tanner, R.I.; Fan, X.-J. *Injection Molding: Integration of Theory and Modeling Methods*; Springer Science & Business Media: Berlin, Germany, 2011.
3. Dimla, D.; Camilotto, M.; Miani, F. Design and optimisation of conformal cooling channels in injection moulding tools. *J. Mater. Process. Technol.* **2005**, *164*, 1294–1300. [CrossRef]
4. Spina, R.; Spekowius, M.; Hopmann, C.J.M. Multiphysics simulation of thermoplastic polymer crystallization. *Mater. Des.* **2016**, *95*, 455–469. [CrossRef]
5. Xu, X.; Sachs, E.; Allen, S.; Cima, M. Designing conformal cooling channels for tooling. In Proceedings of the Solid Freeform Fabrication, Austin, TX, USA, 10–12 August 1998; pp. 131–146.
6. Sachs, E.; Wylonis, E.; Allen, S.; Cima, M.; Guo, H. Production of injection molding tooling with conformal cooling channels using the three dimensional printing process. *Polym. Eng. Sci.* **2000**, *40*, 1232–1247. [CrossRef]
7. Xu, X.; Sachs, E.; Allen, S. The design of conformal cooling channels in injection molding tooling. *Polym. Eng. Sci.* **2001**, *41*, 1265–1279. [CrossRef]
8. Sachs, E.; Allen, S.; Guo, H.; Banos, J.; Cima, M.; Serdy, J.; Brancazio, D. Progress on tooling by 3D printing; conformal cooling, dimensional control, surface finish and hardness. In Proceedings of the Eighth Annual Solid Freeform Fabrication Symposium, Austin, TX, USA, 11–13 August 1997; pp. 11–13.
9. Ferreira, J.; Mateus, A. Studies of rapid soft tooling with conformal cooling channels for plastic injection moulding. *J. Mater. Process. Technol.* **2003**, *142*, 508–516. [CrossRef]
10. Meckley, J.; Edwards, R. A Study on the design and effectiveness of conformal cooling channels in rapid tooling inserts. *Technol. Interface J.* **2009**, *10*, 1–28.
11. Hopkinson, N.; Dickens, P. Conformal cooling and heating channels using laser sintered tools. In Proceedings of the Solid Freeform Fabrication Conference, Austin, TX, USA, 7–9 August 2000; pp. 490–497.
12. Yoo, S. Design of conformal cooling/heating channels for layered tooling. In Proceedings of the ICSMA 2008 International Conference on Smart Manufacturing Application, Gyeonggi-do, Korea, 9–11 April 2008; pp. 126–129.
13. Altaf, K.; Rani, A.M.A.; Raghavan, V.R. Fabrication of circular and Profiled Conformal Cooling Channels in aluminum filled epoxy injection mould tools. In Proceedings of the National Postgraduate Conference (NPC), Kuala Lumpur, Malaysia, 19–20 September 2011; pp. 1–4.
14. Li, C.; Li, C.; Mok, A. Automatic layout design of plastic injection mould cooling system. *Comput.-Aided Des.* **2005**, *37*, 645–662. [CrossRef]
15. Saifullah, A.; Masood, S. Finite element thermal analysis of conformal cooling channels in injection moulding. *Eng. Aust.* **2007**, *1*, 337–341.
16. Saifullah, A.; Masood, S.; Sbarski, I. New cooling channel design for injection moulding. In Proceedings of the World Congress on Engineering, London, UK, 1–3 July 2009.
17. Saifullah, A.B.M.; Masood, S.H. Optimum Cooling Channels Design and Thermal Analysis of an Injection Moulded Plastic Part Mould. *Mat. Sci. Forum* **2007**, *561–565*, 1999–2002.
18. Xu, R.X.; Sachs, E. Rapid thermal cycling with low thermal inertia tools. *Polym. Eng. Sci.* **2009**, *49*, 305–316. [CrossRef]
19. Sun, Y.; Lee, K.; Nee, A. The application of U-shape milled grooves for cooling of injection moulds. *Proc. Inst. Mech. Eng. Part B J. Eng. Manuf.* **2002**, *216*, 1561–1573. [CrossRef]
20. Sun, Y.; Lee, K.; Nee, A. Design and FEM analysis of the milled groove insert method for cooling of plastic injection moulds. *Int. J. Adv. Manuf. Technol.* **2004**, *24*, 715–726. [CrossRef]
21. ó Gloinn, T.; Hayes, C.; Hanniffy, P.; Vaugh, K. FEA simulation of conformal cooling within injection moulds. *Int. J. Manuf. Res.* **2007**, *2*, 162–170. [CrossRef]

22. Au, K.; Yu, K. A scaffolding architecture for conformal cooling design in rapid plastic injection moulding. *Int. J. Adv. Manuf. Technol.* **2007**, *34*, 496–515. [CrossRef]

23. Hsu, F.; Wang, K.; Huang, C.; Chang, R. Investigation on conformal cooling system design in injection molding. *Adv. Prod. Eng. Manag.* **2013**, *8*, 107–115. [CrossRef]

24. Park, H.-S.; Dang, X.-P. Optimization of conformal cooling channels with array of baffles for plastic injection mold. *Int. J. Precis. Eng. Manuf.* **2010**, *11*, 879–890. [CrossRef]

25. Dang, X.-P.; Park, H.-S. Design of U-shape milled groove conformal cooling channels for plastic injection mold. *Int. J. Precis. Eng. Manuf.* **2011**, *12*, 73–84. [CrossRef]

26. Khan, M.; Afaq, S.K.; Khan, N.U.; Ahmad, S. Cycle Time Reduction in Injection Molding Process by Selection of Robust Cooling Channel Design. *ISRN Mech. Eng.* **2014**, *2014*, 968484. [CrossRef]

27. Zink, B.; Kovács, J.G. The effect of limescale on heat transfer in injection molding. *Int. Commun. Heat Mass Transfer.* **2017**, *86*, 101–107. [CrossRef]

28. Wang, Y.; Yu, K.-M.; Wang, C.C.; Zhang, Y. Automatic design of conformal cooling circuits for rapid tooling. *Comput.-Aided Des.* **2011**, *43*, 1001–1010. [CrossRef]

29. Choi, J.H.; Kim, J.S.; Han, E.S.; Park, H.P.; Rhee, B.O. Study on an optimized configuration of conformal cooling channel by branching law. In Proceedings of the ASME 2014 12th Biennial Conference on Engineering Systems Design and Analysis, Copenhagen, Denmark, 25–27 July 2014.

30. Park, H.S.; Pham, N.H. Automatically generating conformal cooling channel design for plastic injection molding. *Ann. DAAAM Proc.* **2007**, 539–541.

31. Park, H.-S.; Pham, N.H. Design of conformal cooling channels for an automotive part. *Int. J. Automot. Technol.* **2009**, *10*, 87–93. [CrossRef]

32. Wang, Y.; Yu, K.-M.; Wang, C.C. Spiral and conformal cooling in plastic injection molding. *Comput.-Aided Des.* **2015**, *63*, 1–11. [CrossRef]

33. Au, K.; Yu, K.; Chiu, W. Visibility-based conformal cooling channel generation for rapid tooling. *Comput.-Aided Des.* **2011**, *43*, 356–373. [CrossRef]

34. Au, K.; Yu, K. Variable Distance Adjustment for Conformal Cooling Channel Design in Rapid Tool. *J. Manuf. Sci. Eng.* **2014**, *136*, 044501. [CrossRef]

35. Agazzi, A.; Sobotka, V.; Le Goff, R.; Jarny, Y. Uniform Cooling and Part Warpage Reduction in Injection Molding Thanks to the Design of an Effective Cooling System. *Key Eng. Mat.* **2013**, *554–557*, 1611–1618. [CrossRef]

36. Mayer, S. *Optimised Mould Temperature Control Procedure Using DMLS*; EOS Whitepaper; EOS GmbH Ltd.: Krailling, Germany, 2005; pp. 1–10.

37. Jahan, S.A. Optimization of Conformal Cooling Channels in 3D Printed Plastic Injection Molds. Master's Thesis, Indiana University Purdue University Indianapolis, Indianapolis, IN, USA, 2016.

38. Jahan, S.A.; El-Mounayri, H. Optimal Conformal Cooling Channels in 3D Printed Dies for Plastic Injection Molding. *Procedia Manuf.* **2016**, *5*, 888–900. [CrossRef]

39. Jahan, S.A.; Wu, T.; Zhang, Y.; El-Mounayri, H.; Tovar, A.; Zhang, J.; Acheson, D.; Nalim, R.; Guo, X.; Lee, W.H. Implementation of conformal cooling & topology optimization in 3D printed stainless steel porous structure injection molds. *Procedia Manuf.* **2016**, *5*, 901–915

40. Jahan, S.A.; Wu, T.; Zhang, Y.; Zhang, J.; Tovar, A.; El-Mounayri, H. Effect of Porosity on Thermal Performance of Plastic Injection Molds Based on Experimental and Numerically Derived Material Properties. In *Mechanics of Additive and Advanced Manufacturing, Volume 9*; Springer: Berlin, Germany, 2018; pp. 55–63.

41. Jahan, S.A.; Wu, T.; Zhang, Y.; Zhang, J.; Tovar, A.; Elmounayri, H. Thermo-mechanical design optimization of conformal cooling channels using design of experiments approach. *Procedia Manuf.* **2017**, *10*, 898–911. [CrossRef]

42. Wu, T.; Jahan, S.A.; Kumaar, P.; Tovar, A.; El-Mounayri, H.; Zhang, Y.; Zhang, J.; Acheson, D.; Brand, K.; Nalim, R. A Framework for Optimizing the Design of Injection Molds with Conformal Cooling for Additive Manufacturing. *Procedia Manuf.* **2015**, *1*, 404–415. [CrossRef]

43. Wu, T.; Jahan, S.A.; Zhang, Y.; Zhang, J.; Elmounayri, H.; Tovar, A. Design optimization of plastic injection tooling for additive manufacturing. *Procedia Manuf.* **2017**, *10*, 923–934. [CrossRef]

44. Design Optimization of Injection Molds with Conformal Cooling for Additive Manufacturing. Available online: https://scholarworks.iupui.edu/handle/1805/9477 (accessed on 1 November 2018).

45. Jahan, S.; El-Mounayri, H.; Tovar, A.; Shin, Y.C. A Framework for Estimating Mold Performance Using Experimental and Numerical Analysis of Injection Mold Tooling Prototypes. In *Mechanics of Additive and Advanced Manufacturing*; Springer: Berlin, Germany, 2019; Volume 8, pp. 71–76.

46. Jauregui-Becker, J.M.; Tosello, G.; Van Houten, F.J.; Hansen, H.N. Performance evaluation of a software engineering tool for automated design of cooling systems in injection moulding. *Procedia CIRP* **2013**, *7*, 270–275. [CrossRef]

MDPI

St. Alban-Anlage 66

4052 Basel

Switzerland

Tel. +41 61 683 77 34

Fax +41 61 302 89 18

www.mdpi.com

*Applied Sciences* Editorial Office

E-mail: applsci@mdpi.com

www.mdpi.com/journal/applsci

www.ingramcontent.com/pod-product-compliance
Lightning Source LLC
Chambersburg PA
CBHW051910210326
41597CB00033B/6092